CHROMATOGRAPHY IN ORGANIC MICROANALYSIS

A LABORATORY GUIDE

CHROMATOGRAPHY IN ORGANIC MICROANALYSIS

A LABORATORY GUIDE

Raphael Ikan

Department of Organic Chemistry
Hebrew University, Jerusalem

ACADEMIC PRESS

A Subsidiary of Harcourt Brace Jovanovich, Publishers

New York   London
Paris   San Diego   San Francisco   São Paulo   Sydney   Tokyo   Toronto

To my wife Yael
and our offspring
Amiran, Ariel, Arnon and Eliana

KETER PUBLISHING HOUSE
Givat Shaul B, Jerusalem

United States Edition published by
ACADEMIC PRESS, INC.
111 Fifth Avenue, New York, New York 10003

ISBN 0-12-370580-0
Library of Congress No. 82-71406

This book has been composed and printed at Keterpress Enterprises, Jerusalem

Contents

Preface

The syllabus of the practical course in organic chemistry at most universities covers the synthesis and qualitative analysis of organic compounds. For some time we have felt that it would be of value to the students, either at the advanced undergraduate or graduate level, to take a comprehensive course in organic analytical chemistry based mainly on chromatographic reactions and separations.

Chemistry is strongly reliant on separation and identification techniques, and indeed the term for chemistry in old Dutch is "scheikunde," literally meaning the art of separation.

The most effective method for complete separation of a complex multicomponent mixture is chromatography (paper, thin-layer, gas-liquid, high-pressure liquid), and the most effective methods of identification are either chemical or spectroscopic [mass spectrometry (MS), nuclear magnetic resonance (NMR), infrared (IR) and ultraviolet (UV)]. Combinations of chromatographic and spectroscopic techniques, such as gas chromatography/MS, are widely used.

During the past decade a new approach has been developed, reaction chromatography, which is a combination of thin-layer and gas-liquid chromatography involving chemical reactions which can be carried out prior to, during, or immediately after the chromatographic separation. Another widely used technique is the pyrolysis of organic compounds, which furnishes information about the structure of unknown compounds.

High-pressure liquid chromatography, which has recently found wide application, has many advantages over other chromatographic techniques and is extensively utilized in this manual.

The success achieved in synthesizing, separating and efficiently identifying minute quantities of compounds by chromatographic methods leads to the solution of problems previously considered insoluble.

The book in its present form includes the following chapters:

1. Microsyntheses on Thin Layers of Silica Gel;
2. Separation of Isomers (Geometrical and Optical);
3. **Reaction Chromatography**;

4. Determination of Food Constituents by Chromatographic Techniques;
5. Forensic Analysis.

Each chapter includes a brief introduction and attempts to supply or recall knowledge in the particular field. Lists of recommended books and reviews for additional study are provided at the end of the chapters.

Typical experiments were selected for each chapter, taking into consideration the following factors: equipment, availability of the starting materials, and the performance time, which is usually very short.

A large number of experiments are described in order to give the instructor a reasonable degree of freedom.

Jerusalem, January, 1982.   Professor RAPHAEL IKAN

Acknowledgements

The author is indebted to his academic colleagues who reviewed the manuscript at different stages of its development and offered constructive criticism: Professor Eli Grushka (Hebrew University of Jerusalem), Bernard Crammer (Bar-Ilan University), Baruch Glattstein (Police Analytical Laboratory, Jerusalem), and to Mrs. Ora Haber (Hebrew University of Jerusalem) for performing many experiments included in the manual.

The following publishers kindly granted permission to reproduce selected material: Marcel Dekker, Inc., The Association of Official Analytical Chemists, American Chemical Society, Springer Verlag, and Elsevier Scientific Publishing Company.

Sincere thanks are due to the staff of Keter Publishing House, particularly Mrs Beth Elon, and to Robert Amoils, who edited the manuscript.

Introduction

The field of organic chemistry has experienced very rapid growth during the past two decades. Enormous strides have been made in pharmaceuticals, plastics, explosives, detergents, agricultural and textile products, petro-chemicals and chemical intermediates. This expansion has led to an ever-increasing emphasis on the analytical aspect of organic chemistry.

The essential challenges facing modern analytical chemistry are to increase the sensitivity and improve the detection limits of analytical processes used to detect and determine low concentrations of organic substances in complex mixtures. This goal was achieved by utilization of modern chromatographic techniques, such as thin-layer, gas-liquid, and high-pressure liquid chromatography.

Chromatography is a physicochemical method of separation in which simple or complex mixtures are separated by distribution between a mobile and a stationary phase. The mobile phase can be a gas or a liquid. Thus, gas chromatography (GC) is divided into gas-liquid chromatography (GLC) and gas-solid chromatography (GSC). Liquid chromatography (LC) is divided into two main types: column chromatography and planar chromatographic methods such as thin-layer chromatography (TLC) and paper chromatography (PC). Column chromatography is subdivided into the following five major types: high-pressure liquid chromatography (HPLC); liquid-solid (or adsorption) chromatography (LSC); liquid-liquid (or partition) chromatography (LLC); bonded-phase chromatography (BPC); ion-exchange chromatography (IEC); and exclusion chromatography (EC). The latter includes gel-permeation chromatography (GPC) and gel-filtration chromatography (GFC). Classification of these chromatographic methods is summarized on the following page.

Since GLC and HPLC have higher separation and detection efficiencies than TLC, their common and distinctive features may be compared.

A sample for GC analysis must be volatile and thermally stable. Samples for LC must be soluble in the mobile phase and differently retarded by the stationary phase. Since much of the work in LC was carried out with *polar* stationary phases (such as silica gel), the use of a *nonpolar* stationary phase is known as "reversed-phase" LC.

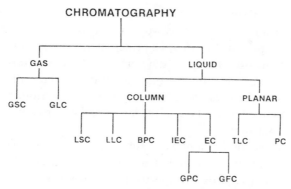

Classification of chromatographic methods.

Another distinction between GC and LC relates to the differences in the physical properties of a gas and a liquid. A liquid is more viscous than a gas, and therefore a high pressure is required for percolating it through a packed HPLC column.

GC and modern HPLC also differ in the particle size of the stationary phase: 125 to 149 μm for the former, and 5 or 10 μm for the latter. The columns used for modern HPLC are short and straight, usually 25 cm in length, as compared to a few meters to hundreds of meters (or longer) in GC.

The mobile phase in GC is usually an inert gas which does not influence the separation. Gases, liquids and solids can be analyzed by GC. The normal range of the molecular weight of gases and compounds is up to 500. LC is applicable to liquids and solids with ionic or covalent bonds. Compounds with molecular weights ranging from 32 to 1,000,000 can be easily and rapidly analyzed. The flame-ionization detector in GC can detect 10^{-11} g and the sensitivity of the electron-capture detector is 10^{-12} g. The column efficiency (number of theoretical plates) of packed GC columns of reasonable lengths is equivalent to 2000–10,000 plates, while that of capillary columns is equivalent to 50,000–100,000 plates per 17 meters. Typical HPLC column efficiency is about 40,000 theoretical plates per meter. Both GLC and HPLC have analytical and preparative capabilities. Quantitative evaluation of chromatograms can be accomplished by a variety of procedures, including computing integrators. In the case of TLC, the chromatoplate spots of compounds are estimated most rapidly by photodensitometric techniques.

Recently, various combinations of TLC and GC techniques have been employed in order to obtain a more complete separation of all the components in complex mixtures. In most cases, the mixture is subjected to preliminary separation by TLC, followed by a complete separation of the preseparated compounds by GC. The principal advantage of the GLC/GC combination stems from the quite different distribution factors which control the separation process. In GC, the separation is principally determined by the volatility of the substances, while in TLC the distribution coefficient is in effect determined by the type of the functional groups present, volatility playing a less important role.

Complete analysis of a complex mixture can be achieved by passing the separate compounds from the outlet of a gas chromatograph into a mass spectrometer which records the spectrum of the individual components. MS is particularly suitable for *combination* with GC. The commonest type of mass spectrometer resolves individual mass numbers up to about 1000; a high-resolution instrument is effective for mass numbers of the order of 10,000 or more.

Computer systems are available to store each mass spectrum and correlate it with the corresponding peak of the gas chromatogram.

There are several techniques in which thin-layer or gas chromatograms may be coupled with *infrared* spectrophotometers.

The chemical identity of compounds separated by TLC or GLC can be confirmed by micro-reactions performed on thin-layer chromatoplates or with GC-effluents. This technique is called *reaction* or *derivation chromatography*.

Substances with high boiling points, such as polymers, can be pyrolyzed, and the more volatile fragments are identified by GC and GC/MS techniques. The advantages of *pyrolysis chromatography* are the short time of analysis, the small amount of sample needed, and the possibility of analyzing both gaseous and liquid products.

The combination and computerization of *chemical* and *chromatographic* methods is a more efficient tool for the identification of organic compounds than either of the two methods individually.

In his book *Philosophy of Sciences*, published in 1838, Andre Marie Ampere made use of the strange term *Cybernetique*, which in its modern concept denotes the multiscient robot or self-organizing machine. It seems that computer-controlled *cybermachines*, such as coupled instruments (GC/MS, TLC/GC, TLC/IR, GC/IR, TLC/GLC/IR), will soon become essential tools in every modern analytical laboratory.

1
Microsyntheses on Thin Layers of Silica Gel

INTRODUCTION

The terms macro, semi-micro and micro, as applied to preparative or synthetic procedures, are differentiated according to the size of the sample taken for synthesis; quantities up to 50 mg are regarded as micro, 50 mg to 1 g as semi-micro, and 1 g or more as macro.

In many laboratories for students of organic chemistry, most syntheses are performed on a macro- and a semi-microscale. The latter technique is of great advantage when the available quantity of the sample is small. However, it has been claimed that in all micropreparative work four limitations must be borne in mind: (a) the reaction should be as free as possible from side ractions; (b) the procedure should involve as few transfer operations as possible; (c) the reaction vessel should be as small as possible; (d) the appropriate method for the isolation of the crude product and its subsequent purification has to be selected.

In order to eliminate these restrictions and further reduce the quantities of the reactants, it was of importance to introduce a *micro*technique for the more common organic preparations.

Typical organic reactions, such as cyclization, dehydration, nitration, oxidation and reduction, are carried out on microscale, using microscope slides covered with a thin layer of silica gel G. This technique involves spotting the sample on the starting line of the slide (covered with a thin layer of silica gel G) and then covering it with a reagent. Reference compounds are also spotted on the starting line. After completion of the reaction, development in a suitable solvent separates the products of the reaction. The R_f value of the original compound, coupled with the chromatographic results for the reaction products and reference compounds, is sufficient for identification of the compound and its reaction products.

In cases where this technique is not suitable, the reagent and the compound can be mixed on a microscale in a small test tube or in a capillary. The crude mixture can then be applied to the chromatoplate.

In many cases (owing to the short reaction period) the reactions on slides do not reach completion, giving rise to a mixture of the original compound and the resulting reaction products.

The inert character of the thin-layer material (usually silica gel) makes it ideally suited for use with strong and corrosive reagents.

This technique provides a great deal of information about the unknown compound and the reaction products with the expenditure of a very small amount of material.

MICROSYNTHESES INVOLVING ELIMINATION OF WATER

Elimination of a molecule of water is characteristic of reactions such as dehydration, cyclization and condensation. These reactions are illustrated by the following microsyntheses.

Microsynthesis of cyclohexene

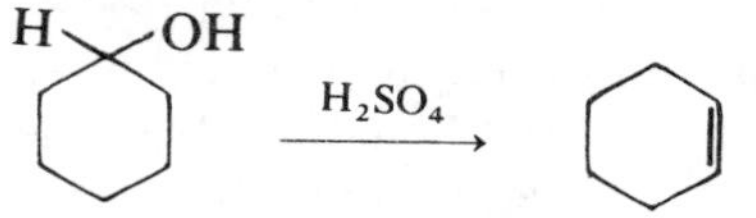

Cyclohexanol   Cyclohexene

The common methods of dehydration are: vapor-phase thermal dehydration in the presence of a catalyst, and liquid-phase dehydration by the action of dehydrating agents such as concentrated sulfuric acid. Cyclohexene is formed by treating cyclohexanol with sulfuric acid at 150°C.

Procedure

A drop of cyclohexanol is placed in a small test tube, a drop of concentrated sulfuric acid is added, and the mixture is heated on a water bath for 5 minutes, diluted with 1 ml of ether and spotted with a micropipette on the starting line of the silica gel-coated microscope slide (10 mm from its rim). After evaporation of the ether, the slide is chromatographed with hexane-ethyl acetate, 9 : 1. After 10 minutes it is removed, air-dried, sprayed with H_2SO_4 (50%) and heated in an oven at 150°C. Cyclohexene appears as a black spot on the upper part of the slide, near the solvent front. Similarly, geraniol may be used instead of cyclohexanol to produce dipentene, the formation of this cyclic terpene proceeding via dehydration and subsequent cyclization reactions.

Microsynthesis of anthraquinone

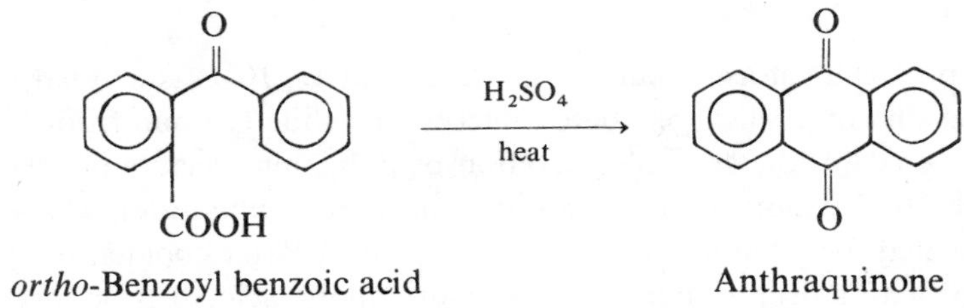

ortho-Benzoyl benzoic acid   Anthraquinone

The formation of anthraquinone from ortho-benzoyl benzoic acid is a typical cyclization reaction (intramolecular loss of water) and is effected by concentrated sulfuric acid at 150°C. Polyphosphoric acid (PPA) is also a convenient reagent for carrying out such cyclizations.

Procedure

A drop of benzene solution of ortho-benzoyl benzoic acid (5%) is spotted with a micropipette on the starting line of the silica gel-coated microscope slide (10 mm from its rim), covered with a drop of concentrated sulfuric acid, and allowed to react for 5 minutes. The slide is then heated for 10 minutes in an oven at 150°C, cooled, a drop of chloroform solution of anthraquinone is spotted as a reference compound, and it is then chromatographed with hexane-ethyl acetate, 9:1. After 10 minutes the slide is air-dried and sprayed with a solution of 2,4-dinitrophenylhydrazine (DNP) (prepared by dissolving 1 g of 2,4-DNP in 5 ml of concentrated sulfuric acid and 7.5 ml of water, and diluting with 25 ml of ethanol) and heated at 150°C. Anthraquinone appears as a brown spot on a yellow background, R_f 0.35.

Microsynthesis of cinnamic acid

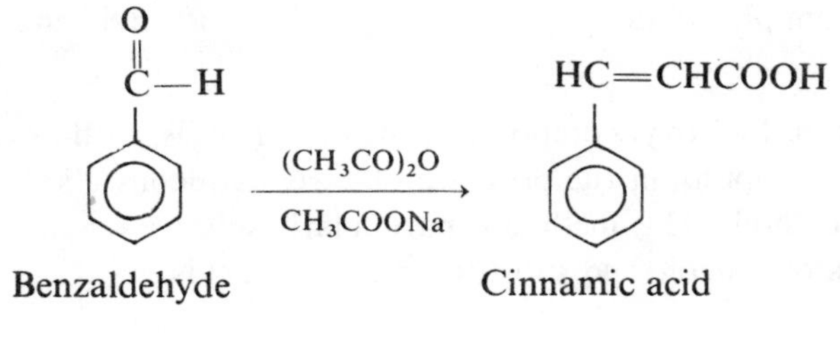

Benzaldehyde   Cinnamic acid

A base-catalyzed, aldol-type condensation of aromatic aldehydes with acetic anhydride and its corresponding sodium salt is called the Perkin reaction. The products of such condensations are cinnamic acids.

Procedure

A drop of chloroform solution of benzaldehyde (10%) is spotted on the starting line of a silica gel-coated microscope slide (10 mm from its rim) and covered with a drop of a solution of potassium acetate in acetic anhydride (0.12 g potassium acetate in 1 ml acetic anhydride). The slide is then heated for 10 minutes in an oven at 170°C. After cooling, the spot is covered with a drop of concentrated hydrochloric acid. A drop of chloroform solution of cinnamic acid is spotted as a reference compound, and the slide is then developed with ethanol-ammonia-water, 100:16:12. After evaporation of the solvents, the slide is sprayed with a solution of bromocresol green [prepared by dissolving 0.04 g of the reagent in 100 ml ethanol (96%), and adding sodium hydroxide (0.1 N) dropwise until a blue color appears] and heated in an oven at 130°C. Cinnamic acid forms a white spot on a blue background, R_f 0.84.

MICROSYNTHESES INVOLVING ACYLATIONS

The high reactivity of acid halides and anhydrides makes them effective reagents for introducing an acyl group into compounds having an active hydrogen atom.

Microsynthesis of phenyl benzoate

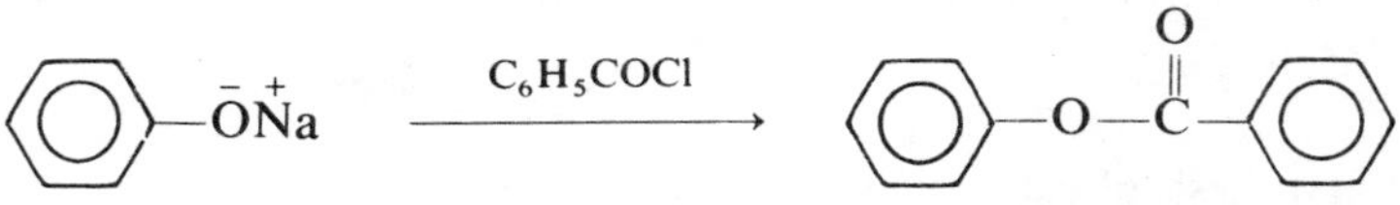

Sodium phenolate   Phenyl benzoate

Acylation of hydroxyl compounds, such as phenols, with aroyl chlorides is carried out rapidly in the presence of excess aqueous alkali (Schotten–Baumann method). This method is particularly suitable for the preparation of the esters of benzoic and substituted benzoic acids.

Procedure

A drop of sodium phenolate solution [prepared by dissolving 0.2 g phenol in 4 ml of sodium hydroxide solution (4%)] is spotted with a micropipette on the starting line of the silica gel-coated microscope slide (10 mm from its rim) and covered with a drop of benzoyl chloride in benzene (2%). After a few minutes, a drop of phenyl benzoate in benzene is spotted as a reference compound, and the slide is chromatographed with benzene-ethyl acetate, 9:1. After allowing a few minutes for development, the slide is air-dried and sprayed with sulfuric acid solution (50%). When the slide is heated at 150°C in an oven phenyl benzoate appears as a pink spot, R_f 0.80.

Microsynthesis of acetophenone

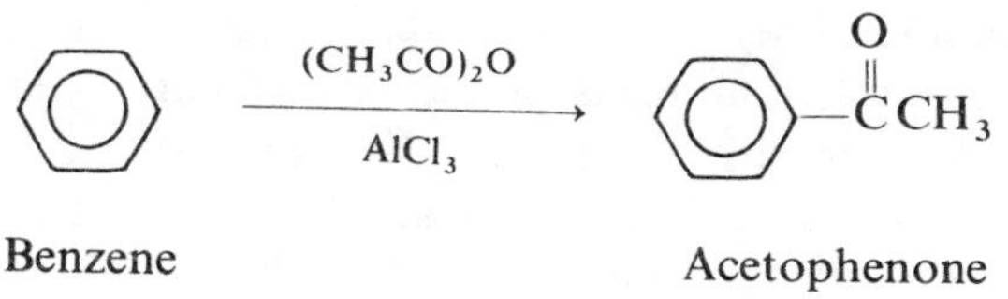

Benzene   Acetophenone

Aliphatic and aromatic acid chlorides in the presence of Lewis acids, such as, for example, aluminum chloride, react with aromatic compounds to furnish alkyl aryl and diaryl ketones.

Acylation of aromatic compounds by means of the Friedel–Crafts reaction is an important synthetic route for the preparation of aromatic ketones.

Procedure

A drop of benzene is placed on the starting line of the silica gel-coated microscope slide (10 mm from its rim) and covered with a few milligrams of anhydrous aluminum chloride, followed by a drop of acetic anhydride. After a few minutes the slide is heated for 10 minutes at 80–90°C in an oven. After cooling, the spot is acidified with concentrated hydrochloric acid. A drop of acetophenone in chloroform is spotted as a reference compound, and the slide is developed with benzene-ethyl acetate, 9:1. The air-dried slide is sprayed with a solution of 2,4-dinitrophenylhydrazine. Acetophenone forms an orange spot, R_f 0.60.

Microsynthesis of acetanilide

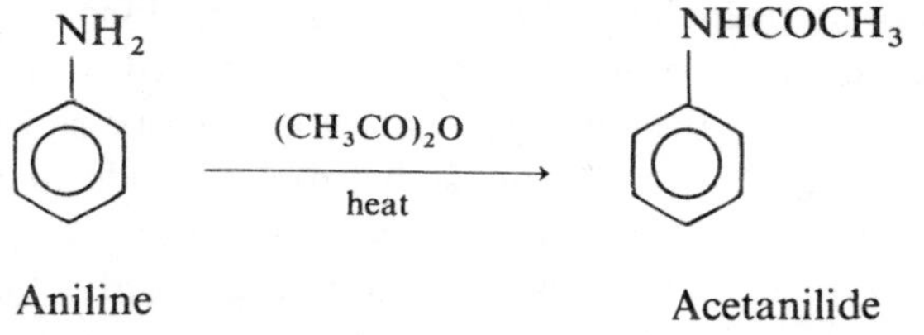

NH$_2$ $\xrightarrow[\text{heat}]{(CH_3CO)_2O}$ NHCOCH$_3$

Aniline   Acetanilide

The amino group of arylamines often is protected by acetylation to reduce their susceptibility to oxidative degradation and to moderate their reactivity in electrophilic substitutions.

Procedure

A drop of a solution consisting of aniline (0.2 ml), concentrated hydrochloric acid (0.2 ml) and water (5 ml) is placed on the starting line of the silica gel-coated microscope slide (10 mm from its rim); after drying, the spot is covered with a drop of acetic anhydride and allowed to react for a few minutes. A drop of reference solution of acetanilide is also spotted on the starting line. The chromatogram is developed with benzene-ethyl acetate, 2:1. After evaporation of the solvents, the slide is sprayed with potassium dichromate solution [2 g potassium dichromate in 50 ml sulfuric acid solution (40%)]. Heating in an oven at 70–80°C reveals acetanilide as a blue spot on a yellow background, R_f 0.23.

MICROSYNTHESES INVOLVING OXIDATIONS

Oxidations on a microscale are illustrated by the following reactions: oxidation of primary alcohol to aldehyde using CrO_3; oxidation of aldehyde to acid using H_2O_2; and oxidation of aldehyde to acid using $KMnO_4$.

Microsynthesis of citral

Many oxidizing agents are available for oxidizing primary alcohols to aldehydes. The most commonly used general oxidizing agent is chromium trioxide, CrO_3, also known as chromic anhydride, which is reduced to Cr(III).

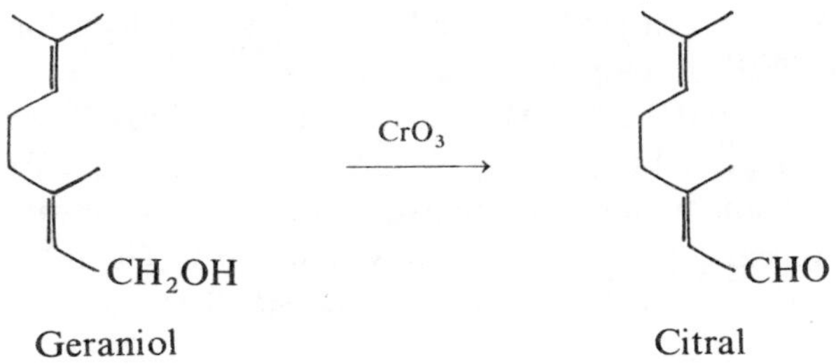

Geraniol   Citral

Procedure

A drop of ethereal solution of geraniol (5%) is spotted with a micropipette on the starting line of the silica gel-coated microscope slide (10 mm from its rim), covered with a drop of 3% solution of chromic anhydride in glacial acetic acid and allowed to react for a few minutes; a drop of citral in ether is applied as a reference. The slide is developed within 10 minutes with hexane-ethyl acetate, 9:1. It is then air-dried and sprayed with a solution of 2,4-dinitrophenylhydrazine. Citral appears as a single orange spot, R_f 0.38.

Microsynthesis of geranic acid

Citral   Geranic acid

Aldehydes are readily oxidized to carboxylic acids. The oxidizing agents commonly used include Ag_2O, H_2O_2, CH_3CO_3H, $KMnO_4$ and CrO_3.

In the following experiment citral is oxidized to geranic acid by means of hydrogen peroxide and exposure to UV light.

Procedure

A drop of ethereal solution of citral (5%) is spotted on the starting line of the silica gel-coated microscope slide (10 mm from its rim), followed by a drop of 30% hydrogen peroxide, and exposed to UV light for 10 minutes. The slide is then developed for 30 minutes with a mixture of ethanol-ammonia-water, 100:12:16. After air-drying, it is sprayed with a solution of bromocresol green [prepared by dissolving 0.04 g of the reagent in 100 ml ethanol (96%), and adding sodium hydroxide (0.1 N) dropwise until a blue coloration appears] and heated in an oven at 130°C. Geranic acid appears as a blue spot on a pale yellow background, R_f 0.63.

Microsynthesis of heptanoic acid

$$CH_3(CH_2)_5CHO \xrightarrow[H^+]{KMnO_4} CH_3(CH_2)_5COOH$$

n-Heptaldehyde n-Heptanoic acid

In the following procedure n-heptaldehyde is oxidized to n-heptanoic acid using potassium permanganate in dilute sulfuric acid as the oxidizing agent.

Procedure

A drop of ethereal solution of n-heptaldehyde (5%) is spotted on the starting line of the silica gel-coated microscope slide (10 mm from its rim), followed by a drop of 1% solution of potassium permanganate in 10% sulfuric acid, and allowed to stand at room temperature for 2 minutes. Heptaldehyde in chloroform is spotted as a reference compound. The slide is developed for 30 minutes with ethanol-ammonia-water, 100:12:16. The part of the slide that was spotted with heptaldehyde is sprayed with 2,4-dinitrophenylhydrazine solution and the part spotted with oxidized heptaldehyde is sprayed with bromocresol green (for preparation, see previous experiment). The slide is then heated in an oven at 150°C. n-Heptanoic acid appears as a pale yellow spot on a blue background, R_f 0.20; heptaldehyde appears as a pale brown spot on a yellow background, R_f 0.64.

Microsynthesis of di-β-naphthol

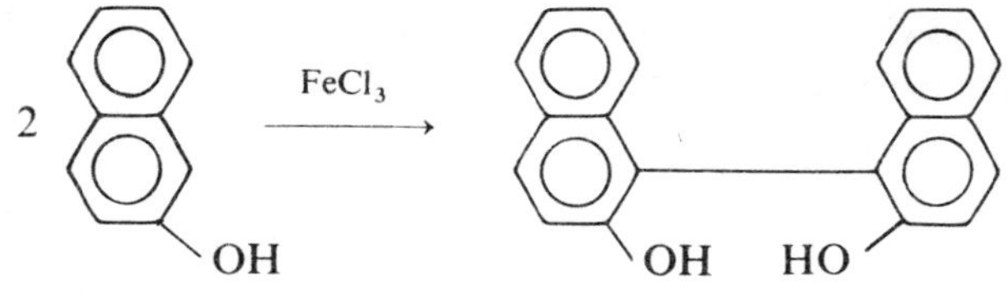

β-Naphthol   Di-β-Naphthol

β-Naphthol is oxidized to di-β-naphthol by aqueous ferric chloride solution.

Procedure

A drop of ethanolic solution of β-naphthol (0.15 g β-naphthol in 6 ml ethanol) is placed on the starting line of the silica gel-coated microscope slide (10 mm from its rim) and covered with a drop of ferric chloride solution (0.3 g $FeCl_3$ in 0.6 ml water). The slide is heated in an oven for 5 minutes at 105°C. A drop of di-β-naphthol in chloroform is placed on the starting line and the slide developed with benzene-ethyl acetate, 9:1. The air-dried slide is sprayed with potassium dichromate reagent (see microsynthesis of acetanilide). Di-β-Naphthol appears as a gray spot on a yellow background, R_f 0.37.

MICROSYNTHESES INVOLVING REDUCTIONS

Reduction of organic compounds is usually carried out with chemical reagents or with hydrogen in the presence of a catalyst, either in the liquid or gaseous phase. Such reductions are illustrated by the following reactions.

Microsynthesis of geraniol

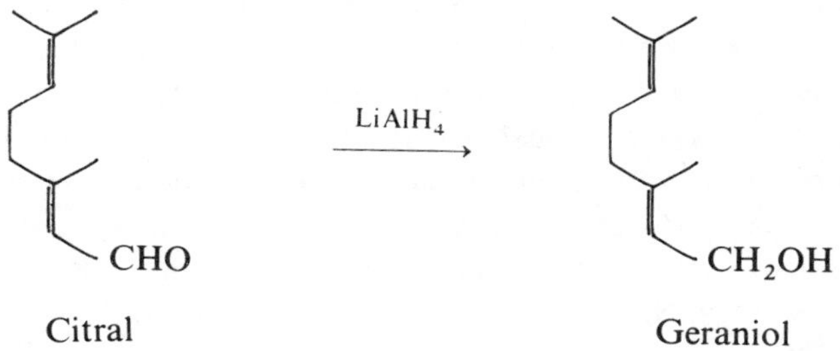

CHO	LiAlH$_4$	CH$_2$OH
Citral		Geraniol

Aldehydes are easily reduced to the corresponding primary alcohols. Many different reducing agents are employed for this purpose. Complex metal hydrides, such as lithium aluminum hydride, are widely used as they are extremely powerful reducing agents.

Procedure

A drop of ethereal solution of citral (5%) is spotted on the starting line of the silica gel-coated microscope slide (10 mm from its rim), covered with a drop of lithium aluminum hydride (1% in dry ether) and allowed to react for a few minutes; a drop of geraniol in ether is also spotted on the slide as a reference. The slide is then developed with hexane-ethyl acetate, 9:1. After 8 minutes, it is air-dried and sprayed with a solution of 2,4-dinitro-phenylhydrazine, whereupon the unreduced citral appears as an orange spot, R_f 0.32. The slide is then heated in an oven at 150°C; geraniol appears as a violet-brown spot (below the citral spot), R_f 0.10.

A similar experiment may be performed with carvone which is reduced to carveol.

Microhydrogenation of cyclohexene

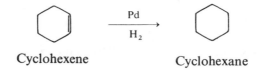

Cyclohexene   Cyclohexane

Procedure

The catalyst is prepared by placing a large spot of 5% palladium or platinum chloride in KCl solution (5%) on the starting line of the silica gel-coated microscope slide (10 mm from its rim), followed by spraying with an alkaline formaldehyde solution (prepared by mixing 20 ml of formaldehyde and 80 ml of 20% aqueous potassium hydroxide). After drying at room temperature, the slide is sprayed with acetic acid (5%) and again dried, at 80°C. A small spot of an unsaturated compound such as cyclohexene is placed on the catalyst and the slide is sprayed with ethyl acetate. Hydrogenation is performed by exposing the slide for several hours to the action of a slow stream of hydrogen in a desiccator containing ethyl acetate. The slide is then dried (cyclohexane is spotted as a reference), developed with a suitable solvent, such as *n*-hexane, and exposed to bromine vapors for the detection of traces of unreduced cyclohexene.

MICROSYNTHESIS INVOLVING NITRATION

Micronitration of phenol

Phenol is nitrated to a mixture of *o*- and *p*-nitrophenols by dilute aqueous nitric acid even at room temperature. The ratio of *ortho-* and *para*-products depends on the concentration of nitric acid in solution.

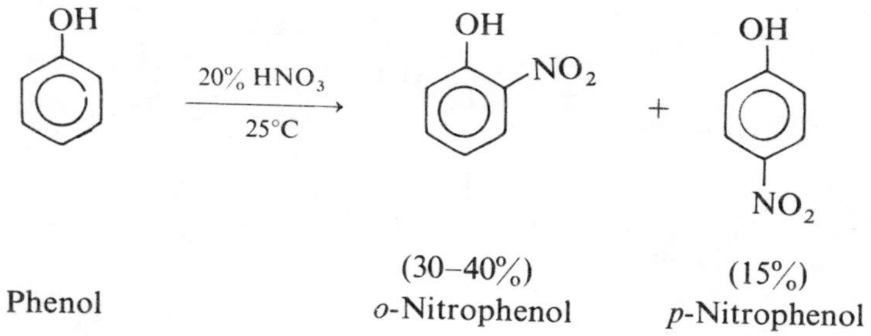

Phenol $\xrightarrow[25°C]{20\% \ HNO_3}$ (30–40%)   +   (15%)
o-Nitrophenol   p-Nitrophenol

The R_f value of the *ortho*-isomer is higher than that of the *para*-isomer; this is probably due to intramolecular hydrogen bonding between the hydroxy group and the nitro group as illustrated below.

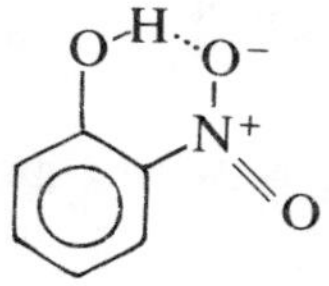

Procedure

A drop of 5% solution of phenol in chloroform is spotted on the starting line of the silica gel-coated microscope slide (10 mm from its rim), covered with a drop of nitric acid-water (2:1) and allowed to react for a few minutes. *Para-* or *ortho*-nitrophenol in chloroform is spotted as a reference. The slide is then chromatographed for ten minutes with benzene-ethyl acetate, 9:1. *Para-* and *ortho*-nitrophenols appear as two distinct yellow spots, their R_f values being 0.24 and 0.77, respectively.

MICROSYNTHESES INVOLVING AZO COUPLING [1]

It is well known that coupling between a diazonium salt and a phenol takes place principally in the *ortho* and *para* positions, as depicted in the following equation:

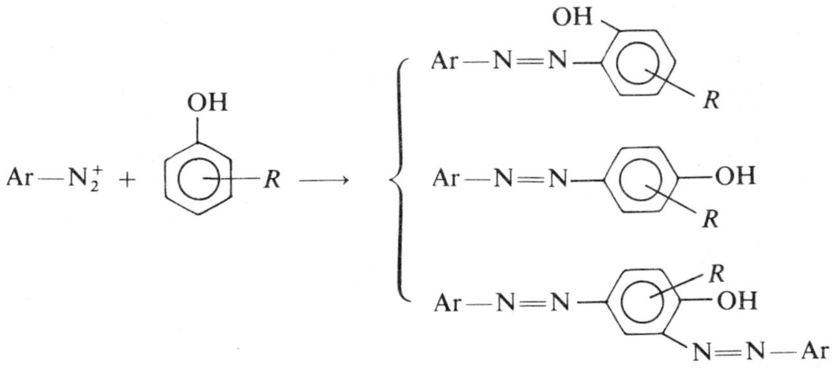

The coupling reaction is carried out on a thin-layer silica gel plate and the reaction products are separated on the same plate. After chromatographic separation, the different spots are eluted with dimethylformamide and the concentrations of various products are determined spectrophotometrically or by measuring the spot size.

Microsynthesis of azo coupling between α-naphthol and 2,5-dimethoxyaniline

Equipment and Materials

Glass plates (20 × 20 cm) or microscope slides coated with silica gel G
Spectrophotometer
2,5-Dimethoxyaniline
α-Naphthol

Procedure

2,5-Dimethoxyanaline (50 mg) is dissolved in a test tube containing 2.5 ml of distilled water and 0.5 ml of concentrated hydrochloric acid. The solution is kept in an ice bath and a crystal of sodium nitrite is added. The diazotization reaction is completed within 5 minutes. α-Naphthol solution is prepared by dissolving 4.5 mg of the compound in 0.5 ml of pyridine. Using an ultramicropipette, 10 μl of the diazonium chloride solution are delivered on the starting line of the thin layer, and then 20 μl of α-naphthol solution are added on the same spot. Each spot thus contains approximately 170 μg of the diazotized amine and 150 μg of α-naphthol, equivalent to a molar ratio of 1:1. The coupling reaction takes place immediately on the plate, as indicated by the color change.

Chromatographic separation of the products

The thin-layer plate is left to stand in the air for 2 minutes in order to allow the coupling reactions to proceed as far as possible. The plate is then developed using benzene-ethyl acetate (2:1). The main product (compound II), which is dark red, readily separates from the other components. The *ortho*-isomer (compound I) appears violet and the *bis*-azo material (compound III) appears blue-black on the plate. The latter two spots tend to overlap, though the violet compound stays at the front.

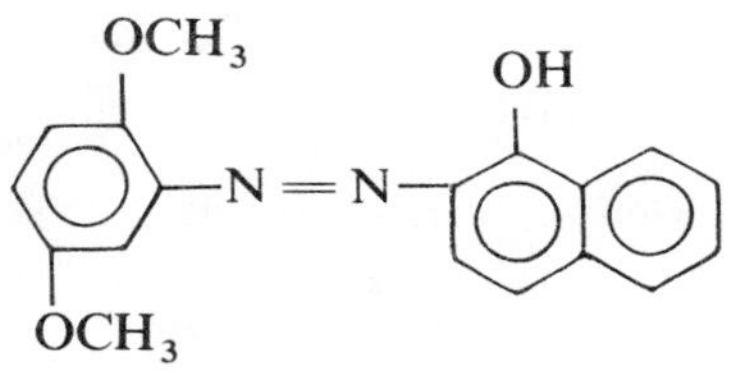

(I) 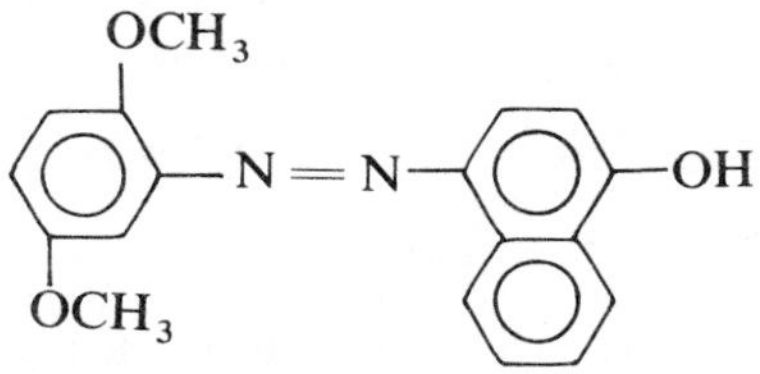   (II)

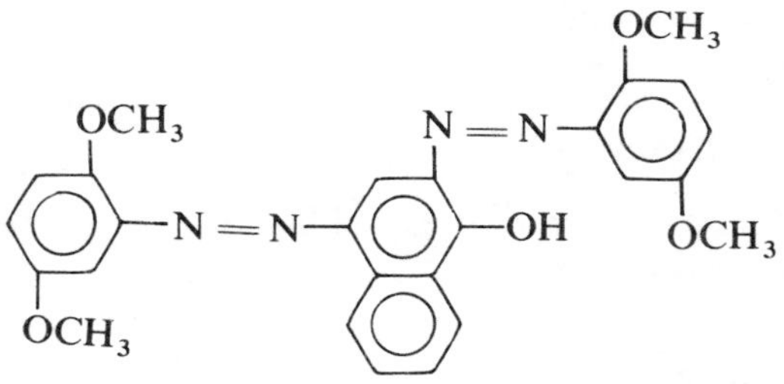

(III)

Estimation of the relative yields

A rough estimate of the comparative yields of the coupling reaction can be made by mere observation of the thin-layer plate. In order to obtain more accurate values, the colored spots are removed from the plate as follows: the plate is dried with a stream of warm air; the respective colored spots are scratched and removed into separate beakers; a few milliliters of dimethylformamide are added with stirring to extract the azo compounds. The solutions are filtered into separate 10-ml volumetric flasks, diluted to the mark and mixed. The absorbance of the solutions in the spectrophotometer is read against a DMF blank.

Micro-multiple synthesis of azo dyes [2]

Simultaneous preparation of several compounds with analogous structures using a general chemical reaction is a common practice in synthetic organic chemistry. In the dyeing industry, multiple syntheses are usually carried out when it is desired to produce a compound which meets specific requirements, such as color, shade, light-fastness, etc. The processes for the separation, isolation and purification of the desired products are tedious and time consuming.

In the following experiment a series of aromatic amines are diazotized according to the equation:

$$ArNH_2 + HNO_2 \longrightarrow Ar-N_2^+ + OH^- + H_2O$$

The diazonium salts are then treated with α-naphthol to form azo dyes:

$$Ar-N_2^+ + C_{10}H_7(OH) \longrightarrow Ar-N{=}N-C_{10}H_6(OH) + H^+$$

It is obvious that the coupling reaction gives rise to more than one product. Using a 20 × 20 cm thin-layer plate, the diazotization of 5 amines, followed by the coupling with α-naphthol on the plate and observation of the resulting products, is accomplished in about 2 hours.

Equipment and Materials

Plates (20 × 20 cm) coated with silica gel G
Microsyringe, graduated in μl
Aromatic amines: *p*-anisidine, *o*-anisidine, *m*-anisidine; *o*-chloroaniline; 2,5-dimethoxyaniline
α-Naphthol

Procedure

Spot application
   Aromatic amines (200 mg each) are dissolved in methanol in separate 5-ml volumetric flasks and diluted to volume. Using a microsyringe, a spot of each of the 5 compounds is applied along a line 35 mm above the bottom of the thin-layer plate, the spots being spaced approximately 30 mm apart.

Diazotization
   Diazotizing solution is freshly prepared by dissolving 500 mg of sodium nitrite in 50 ml of 1 *N* HCl. The thin-layer plate is covered very carefully

with 2 clean glass slabs in such a manner as to expose only the arc (a band of 10 to 15 mm width) where the spots are located. The diazotizing solution is sprayed onto the exposed band until it is well saturated. The plate is left at room temperature for 5 minutes. The glass slabs are carefully removed and the plate is heated at 105°C for 5 minutes in an oven to destroy the excess nitrous acid.

Formation and separation of dyes

Two clean glass slabs are carefully placed on the above thin-layer plate in such a manner as to leave only the area containing the diazonium chlorides exposed. A 5% methanolic solution of α-naphthol is sprayed onto this area and the plate is dried with a stream of warm air. The glass slabs are removed and the thin-layer plate is placed in a chromatographic chamber containing benzene-ethyl acetate (2:1). The plate is developed up to 15 cm above the starting line containing the dyes. As the solvent front moves upward, the colored spots are separated. Development time is about 1 hour.

REFERENCES

1. Fono, A., Sapse, A. M. and Ma, T. S., "Organic Synthesis in the Microgram Scale. I. Investigation of Azo Coupling on the Thin-Layer Silica Gel Plate," *Mikrochim. Acta*, 1098 (1965).
2. Marcus, B. J., Fono, A. and Ma, T. S., "Organic Synthesis in the Microgram Scale. II. A Simple and Rapid Technique for Multiple Synthesis," *Mikrochim. Acta*, 960 (1966).

RECOMMENDED READING

Heftmann, E., *Chromatography*, Van Nostrand Reinhold Co., 1975.
Kirchner, J. G., *Thin Layer Chromatography*, John Wiley, 1978.
Ma, T. S., *Microscale Manipulations in Chemistry*, John Wiley, 1976.
Ma, T. S., *Organic Functional Group Analysis by Micro and Semimicro Methods*, John Wiley, 1964.
Stahl, E., *Thin Layer Chromatography*, Springer Verlag, 1972.
Touchstone, J. C. and Dobbins, M. F., *Practice of Thin Layer Chromatography*, John Wiley, 1978.

2
Separation of Isomers (Geometrical, Optical) and Other Complexes

INTRODUCTION

Many natural mixtures, such as lipids, are readily separated into various components, but these are frequently not single compounds but groups of compounds differing in chain-length, degree of unsaturation, position of functional groups and stereochemistry.

Progress has been made recently in facilitating the separation of some of these groups of closely related compounds by subjecting them to chromatography and adsorbents which have been impregnated with compounds having the ability to complex preferentially with or to interact with specific functional groups.

The nature of the bonding in complexes of unsaturated compounds with silver (and other transition metals with nearly filled d-orbitals) is represented as a resonance hybrid:

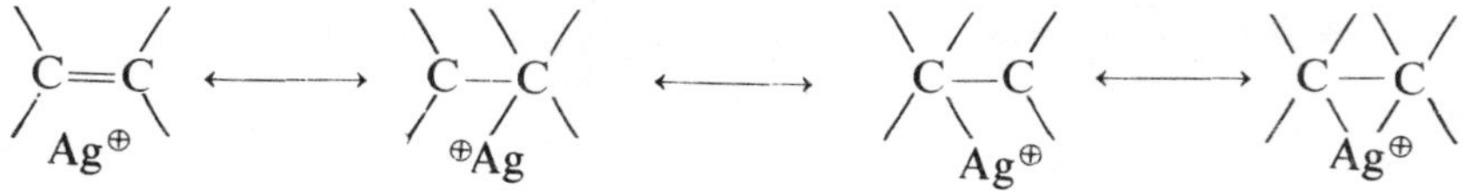

$$\underset{Ag^{\oplus}}{C=C} \longleftrightarrow \underset{^{\oplus}Ag}{C-C} \longleftrightarrow \underset{Ag^{\oplus}}{C-C} \longleftrightarrow \underset{Ag^{\oplus}}{C-C}$$

This is a simple π-complex in which only deformation of the π-orbitals of the olefin is involved.

Other complexing agents used in TLC are boric and phenylboronic acids which form six-membered cyclic derivatives:

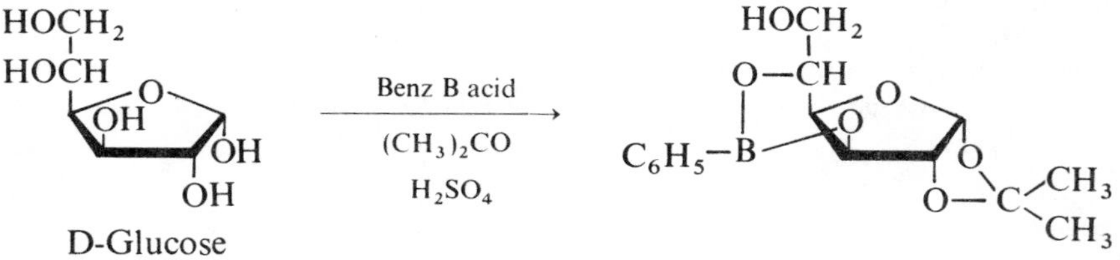

$$
\begin{array}{ccc}
\text{HOCH}_2 & & \text{HOCH}_2 \\
\text{HOCH} & \xrightarrow[\substack{(CH_3)_2CO \\ H_2SO_4}]{\text{Benz B acid}} & \text{O---CH} \\
\text{D-Glucose} & & C_6H_5-B
\end{array}
$$

In the presence of acetone, an acetonide is also formed.

For the separation of polycyclic aromatic hydrocarbons (which are electron donors), thin layers impregnated with picric acid, 2,4,7-trinitrofluorenone, urea, caffeine, 1,3,5-trinitrobenzene and tetracyanoethylene (which are electron acceptors) are used. The interaction of electron-donors and electron-acceptors forms complexes which are visible under UV light.

Many chromatographic techniques have been developed recently for resolving enantiomeric mixtures.

The separation of enantiomeric amino acids (described in this chapter) is based on their conversion into binary diastereomeric Cu(II) complexes, which are resolved by HPLC.

SEPARATION OF POLYUNSATURATED FATTY ACIDS BY ARGENTATION TLC AND GLC [1]

Various procedures are available for the separation of fatty acid methyl esters. The most common is GLC which gives excellent quantitative data but does not lend itself readily to the recovery of appreciable amounts of the esters.

Silver nitrate TLC is used extensively to fractionate methyl esters according to their degree of unsaturation.

The following method makes possible the separation, on the same plate, of samples containing up to 6 double bonds, which are then isolated and analyzed by GLC.

Procedure

Thin-layer plates (5 × 20 cm) are coated with a 0.25 mm-thick slurry of silica gel-silver nitrate (35 ml of 4% w/v $AgNO_3$ to 15 g of silica gel G). The plates are allowed to air-dry for ca. 45 minutes and are then activated for 30 minutes at 110°C and stored in a light-tight, dessicated container. An equimolar mixture of fatty acid methyl esters [$C_{18:0}$, $C_{18:1}$, $C_{18:2}$, $C_{18:3}$, $C_{20:4}$, $C_{20:5}$, $C_{22:6}$ (about 950 μg)] is applied to each plate in a narrow band, and the plates are developed twice in a solvent system consisting of hexane-diethyl ether-acetic acid (94:4:2).

The plate is dried under a stream of nitrogen, sprayed lightly with a 0.1% solution of 4′,5′-dibromofluorescein in isopropanol, dried under nitrogen, and placed in a tank saturated with NH_4OH vapors. After ca. 5 minutes, pink spots appear on a yellow background, which are very pronounced when viewed under UV light. Individual bands are scraped into a 7-ml tube, and a known amount of methyl arachidate ($C_{20:0}$) is added to each tube (as internal standard) to determine the recovery of each methyl ester. Methanol (2 ml) containing 1% acetic acid is added to each tube which is sealed tightly and placed in a boiling water bath for 5 minutes. After cooling and centrifugation, the methanol layer is removed and the silica gel reextracted with the acetic acid-methanol solution. The extracts are combined, the solvents evaporated under a stream of nitrogen, and the residue is injected into a gas chromatograph equipped with a stainless steel or glass column packed with 5% diethylene glycol succinate (DEGS) or any other column used for ester separation.

SEPARATION OF *CIS* AND *TRANS* MONOUNSATURATED FATTY ACIDS BY HPLC [2]

During partial hydrogenation of oils, substantial amounts of positional and geometric fatty acids isomers are formed.

Reports on the metabolic and physiological effects of *cis* and *trans* fatty acids are conflicting and the separation of the isomers is of prime importance in food control.

The following HPLC procedure describes the separation of all positional and geometric isomers of the *trans* $C_{18:1}$ and *cis* $C_{18:1}$ groups.

Procedure

Preparation of silver nitrate-treated silica

Silver nitrate-treated silica is prepared by suspending 10 g Spherisorb S5W (spherical porous silica, 5 μm) in 100 ml of water in which are dissolved varying amounts of silver nitrate (e.g., 2.0 g for a 20% column). The suspension is dried on a rotary evaporator under vacuum at 60°C to a thick paste which is further dried in an oven at 130°C for 2 hours at atmospheric pressure. The cake is broken up and the powder is activated in an oven at 150°C under vacuum for 2 hours. Silver nitrate (2%)-impregnated Partisil 20 is prepared in a similar way.

Packing of column

A slurry of impregnated silica in dioxane is packed into a stainless steel column (25 cm long, 5 mm i.d.) by means of a pump using hexane as the pressure liquid (300 bar). After keeping the column pressurized for ca.

15 minutes, the pressure is lowered to ca. 50 bar and the column flushed with 300 ml UV-grade hexane.

Chromatographic conditions

Mobile phase: Mixtures of UV-grade hexane with tetrahydrofuran (THF) are used. The concentration of THF in the mixture is in the region of 1%.

Detection: UV; 205 nm.

Samples

The samples consist of reference compounds (*trans* and *cis*, $\Delta^5-\Delta^{12}$ $C_{18:1}$ fatty acid methyl esters) in hexane or mixtures of methyl esters, 5% in hexane. The mixture of methyl esters can be obtained by hydrolysis and esterification of commercial fats such as margarine.

Separation of *cis-trans* isomers is accomplished using an impregnated Partisil 20 column, whereas the *positional* isomers of the *cis* $C_{18:1}$ and *trans* $C_{18:1}$ groups, such as Δ^7-*trans* and Δ^9-*trans* or Δ^7-*cis* and Δ^9-*cis*, are successfully separated on a silver nitrate-impregnated Spherisorb S5W column.

The column performance tends to decrease after ca. 20 analyses, but can easily be restored by "washing" the column with several injections of pure THF.

CIS-TRANS ISOMERIZATION OF AZOBENZENE

Many azo compounds show *cis-trans* isomerism. The *trans*-isomer is generally the more stable and the activation energy for conversion to the *cis*-isomer is sufficiently low. In the following experiment *trans*-azobenzene is *partially* converted to the *cis*-isomer by photolysis. The activation energy required to reconvert the *cis*-isomer to *trans*-azobenzene is only 23–25 kcal/mole.

Microsynthesis of *cis*-azobenzene

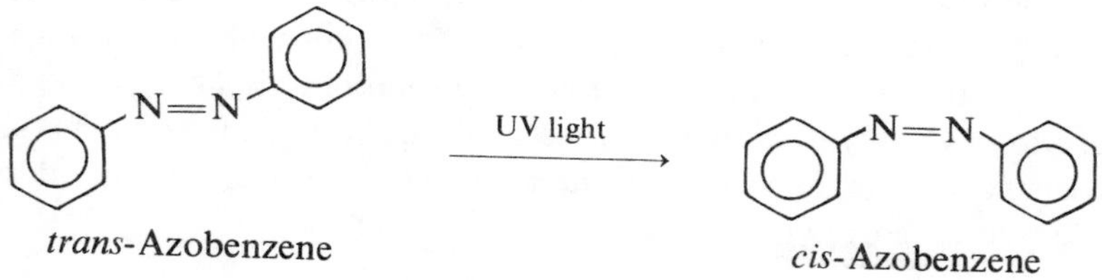

trans-Azobenzene   UV light →   *cis*-Azobenzene

Procedure

A few drops of azobenzene solution (1% in hexane) are spotted on the starting line of the silica gel-coated microscope slide (10 mm from its rim) and then irradiated by UV light for 60 minutes. The slide is developed with hexane-ethyl acetate, 9:1. Air-drying of the slide reveals two yellow spots: the upper spot of the *trans*-isomer, R_f 0.67; and the lower one of the *cis*-isomer, R_f 0.34.

CHROMATOGRAPHY OF POLYCYCLIC AROMATIC HYDROCARBONS ON IMPREGNATED THIN-LAYER PLATES [3]

Aromatic hydrocarbons, as donors of π electrons, are capable of forming donor-acceptor complexes (EDA) with substances having electron-acceptor properties, such as silver nitrate, caffeine, tetracyanoethylene and polynitro substances (e.g., 1,3,5-trinitrobenzene and 2,4,7-trinitrofluorenone). For the separation of polycyclic aromatic hydrocarbons (PAH) from complex mixtures, as encountered, for example, in tobacco smoke, air pollution and pyrolysis studies, the following TLC methods have been used:

1. a plate prepared in advance is impregnated with a complex-forming agent;
2. a complex-forming agent is added to the adsorbent during preparation of the plate;
3. a plate prepared in advance without a complex-forming agent is developed in a system containing the complex-forming agent.

In the following experiment the second method is used for separation of PAH.

Procedure

Hydrocarbons

Test solutions are prepared in methylene chloride, the concentration of polycyclic hydrocarbons being 0.5%. When this concentration cannot be obtained, saturated solutions are used. The solutions are kept in darkness.

The following hydrocarbons can be used: anthracene, pyrene, chrysene, perylene, coronene, fluoranthene, fluorene, acenaphthylene, phenanthrene, triphenylene, benzopyrenes, benzanthracenes, etc.

One microliter of sample solution is used for spotting the plate. Solutions containing more than one hydrocarbon (test mixtures) are prepared by mixing equal volumes of the solutions of the pure substances, and as many microliters are applied to the plate as the number of hydrocarbons in the mixture. For quantitative work (extraction and estimation by UV-spectroscopy) larger amounts of hydrocarbons are needed.

Plates

Glass plates, 20 × 20 cm, covered with a 0.25-mm layer of silica gel or alumina are used. After the adsorbent slurry has been spread and air-dried for half an hour, the plates are activated at 150°C for 3 hours (alumina) or at 120°C for 2 hours (silica gel).

Impregnated layers

The following substances can be used for impregnating the layers: picric acid, 2,4,7-trinitrofluorenone, 1,3,5-trinitrobenzene, tetracyanoethylene, caffeine, silver nitrate. The active layers for the first four compounds are prepared by dissolving each of them in 1 or 2 ml of acetone or alcohol, the resulting solution being added to the amount of water to be used for preparing the layer in the usual way. The complexing agent is thus evenly distributed on the adsorbent.

Caffeine (0.5 g per plate) or silver nitrate (2.5 g per plate) is dissolved in the amount of water necessary for preparing the plates. It is necessary to dissolve caffeine at 50–60°C, taking advantage of its largely enhanced solubility at increased temperature. In order to avoid recrystallization of the caffeine when suspending the adsorbent in the solution, the adsorbent, too, is preheated to the same temperature. The silver-nitrate plates are dried and stored in the dark.

Solvents

Light petroleum and small amounts of polar solvents, such as pyridine (2–5%), ether (1%) and acetic acid (0.04%) in light petroleum, are most useful.

Development

All experiments are run at room temperature. Very often, repeated developments are performed on the plate. The plate is dried for a short time in a stream of air before the next run.

Detection and identification

The spots on the developed chromatograms are located by inspection in UV light (3660 Å). The fluorescence colors on caffeine-impregnated silica-gel plates are much more brilliant than those on ordinary plates.

When nitro compounds are added to the active layer, the fluorescence is not quenched on plates impregnated with silver nitrate. Spraying with concentrated sulfuric acid or nitric acid followed by heating in an oven to 180–200°C, as well as contact with iodine vapor, is also used for detection.

TLC OF CARBOHYDRATES IN THE PRESENCE OF BORIC ACID

The chromatographic separation of polyhydroxy compounds (cardenolides, hydroxy-acids, carbohydrates, polyalcohols, etc.) can be carried out successfully on adsorbents impregnated with boric acid if the component to be isolated contains hydroxyl groups in a position and conformation favorable for the formation of a cyclic boric acid derivative, and if the polarity of this derivative is different from that of the free diol.

Procedure

Preparation of thin layers impregnated with boric acid

For uniform impregnation with boric acid, the plates are placed for a few minutes in the saturated vapor phase of a 1% boric acid solution in methanol. In this solution volatile methyl borate is formed, which is uniformly adsorbed on the thin layer. On removal of the plate from the tank, the methyl ester is decomposed and boric acid remains on the adsorbent. The thin-layer plates impregnated with boric acid can also be prepared by mixing 6 g of silica gel G powder with 15 ml of 0.02 M boric acid solution and coating the plates with this slurry. The plates are then activated at 105°C for 30 minutes in an oven.

Chromatographic conditions

Standard solutions of hexoses and pentoses are prepared by dissolving 20 mg of each sugar in 10 ml of distilled water. Samples of sugar solutions are placed 1.5 cm from the rim of the plate, which is submerged in the developing solvent to a depth of 5 mm. The following developing solvents are used for carbohydrates: (a) chloroform-methanol (3:2); (b) 2-propanol-water (4:1); (c) acetone-water (9:1).

After development the plates are dried, sprayed with aniline-diphenyl-amine reagent [freshly prepared before use by mixing 5 volumes of aniline (12% in acetone), 5 volumes of diphenylamine (2% in acetone) and 2 volumes of concentrated (85%) phosphoric acid], and then heated at 85°C for 10 minutes. The characteristic colors of various sugars after spraying with the chromogen are as follows: lactose—blue-violet; sucrose—lilac; galactose—gray-green; fructose—red-scarlet; glucose—gray-green; arabinose—bright blue; xylose—bright blue; rhamnose—pale green.

Separation of carbohydrates on nonimpregnated thin layers

Sugars can be chromatographed on *non*impregnated silica gel thin layers and developed with a solvent system such as methyl ethyl ketone-2-propanol-acetone-0.5 M H_3BO_3 (40:40:14:16), where boric acid is in the developing system, or a solvent system such as ethyl acetate-2-propanol-water (60:30:10) containing 0.5% phenylboronic acid. After the development, the plates are dried and sprayed with aniline-diphenylamine reagent.

SEPARATION OF THE OPTICAL ENANTIOMERS OF AMINO ACIDS BY REVERSED-PHASE HPLC [4]

Many techniques are available for the separation of enantiomeric amino acids on account of their biological importance and significance in determining the extent of racemization in protein and peptide synthesis.

Enantiomers can be resolved by conversion into diastereomers which are then separated by GLC and LC, or by use of chiral stationary or mobile phases.

In the following procedure the chiral reagent is either a Cu(II) or Zn(II) complex of L-aspartyl-cyclohexylamide.

The mechanism of separation involves the formation of a binary diastereomeric Cu(II) complex. This complex contains one aspartyl unit and one (either L or D) amino acid unit:

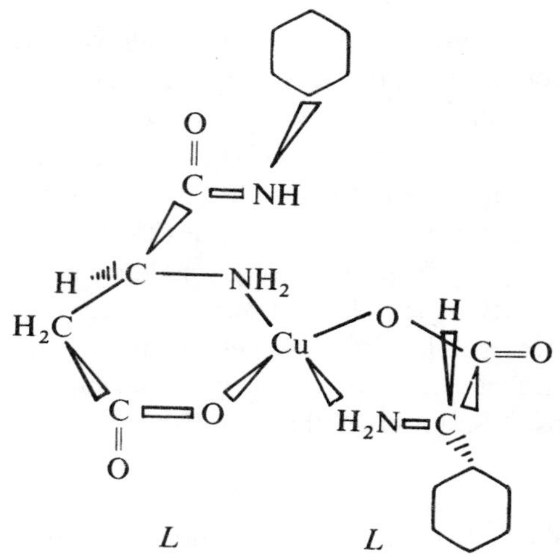

Configuration of the binary complexes.

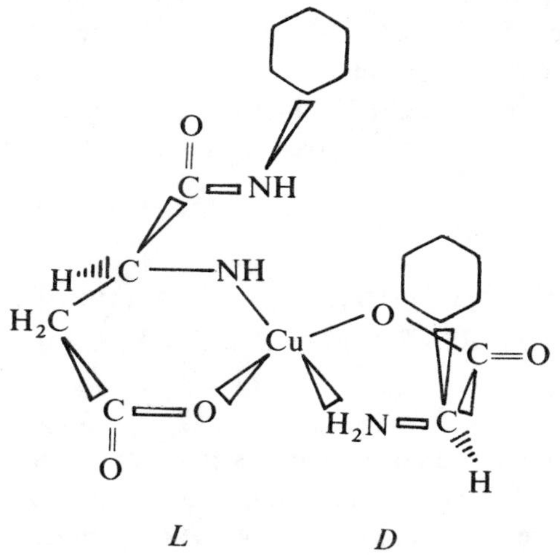

The two diastereomeric complexes (I and II) formed differ markedly in their conformation. In the *DL* complex the amino acid side chain is in close proximity to the cyclohexylamide moiety, thus forming a hydrophobic cluster. In the *LL* complex the amino acid side chain cannot attain this conformation. The two diastereomeric complexes are thus separated on a reversed-phase HPLC column. Detection is accomplished by the adsorbance at 230 nm of the five-membered ring formed by the amino acid in the binary Cu(II) complex.

In all cases the *L*-enantiomers are eluted first, followed by the *D*-enantiomers.

The following amino acids can be separated: proline, valine, cysteine, methionine, isoleucine, tyrosine, dopa, ethionine, phenylalanine, tyrosine.

Equipment and Materials

Column: stainless steel (25 cm long, 4.1 mm i.d.) packed with ODS bonded to Partisil 10.
Mobile phase: aqueous solution of *L*-aspartyl-cyclohexylamide (6×10^{-4} *M*) and $CuCl_2$ (3×10^{-4} *M*); flow rate—2 ml/minute.
Detection: UV, 230 nm.

Preparation of L-*Aspartyl Cyclohexylamide*

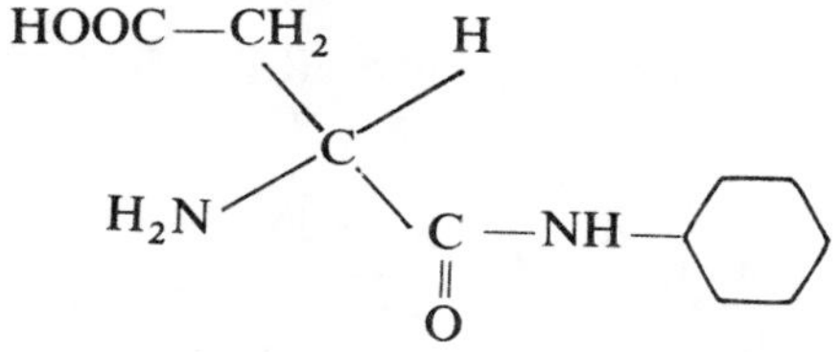

N-Carbobenzoxy-*L*-aspartic acid (Z-Asp(OBzl)OH) (3.7 g) is dissolved in dimethyl formamide (DMF) (25 ml); *N*-methylmorpholine (1.13 ml) is added and the mixture is cooled to $-15°C$. Isobutylchloroformate (1.26 ml) is added and after 2 minutes freshly distilled cyclohexyl amine (cHex) (1.15 ml) is added. The mixture is kept for 1 hour at $-15°C$ and then at $0°C$, and potassium bicarbonate (15 ml, 2*M*) is added with vigorous stirring. After half an hour a solution of NaCl (25%, 100 ml) is added. The white precipitate is collected by filtration and washed with water until pH 7 is reached. Yield, 85%; m.p. 139–141°C.

To a solution of Z-Asp(OBzl)-cHex (3 g) in acetic acid (90%, 30 ml), Pd/C (10%, 300 mg) is added and the mixture hydrogenated at room temperature (5 atmospheres). The catalyst is recovered by filtration. The residue may be crystallized from water. Yield, 86%; m.p. 235°C.

RESOLUTION OF AMINO ACID ENANTIOMERS BY GLC [5]

The resolution of optical enantiomers requires the use of a nonvolatile optically active (chiral) stationary phase coated on an inert support. Vapors of the α-amino acid derivatives, usually the N-acyl derivatives (defined as the "solutes"), undergo resolution as they are carried along the column by an inert gas which serves as a mobile phase, by interacting with the liquid stationary phase (defined as the "solvent").

In the present system solvent and solute are secondary amides which interact principally by hydrogen bonding. The weak bonding allows the solute to reach equilibrium with the solvent as the enantiomeric mixture of the solutes passes along the column. The enantiomeric mixture becomes richer in one of the enantiomers. The solvent-solute interaction is of a diastereomeric nature. Thus if R is the optical activity of the solvent and R' and and S' are the enantiomeric pair to be resolved of the solute, the equilibrium established is then given as:

$$R + R' \; \overset{k_1}{\rightleftharpoons} \; RR'$$
$$R + S' \; \overset{k_2}{\rightleftharpoons} \; RS'$$

where

$$k_1 \neq k_2$$

The equilibrium constants k_1 and k_2 are different, hence the R' and S' enantiomers elute from the column at different retention times.

In the following experiment DL-trifluoroacetyl alanine methyl ester is resolved by GLC into D and L enantiomers on a chiral stationary phase which is a nonvolatile amino acid derivative (lauroyl-L-valyl-$tert$-butyl-amide).

Materials

Lauroyl-L-valyl-$tert$-butylamide:

$$L\text{-}CH_3(CH_2)_{10}CONHCH[CH(CH_3)_2]CONHC(CH_3)_3$$

can be obtained from Miles-Yeda Co., Rehovot, Israel.

Preparation of DL-*Trifluoroacetyl-Alanine-Methyl Ester Hydrochloride*
$CF_3CO(NH_2)COOCH_3 \cdot HCl$

DL-Alanine (500 mg) is introduced into an ampoule containing 2 ml 1.25 N methanolic hydrochloric acid (methanolic hydrochloric acid is prepared by bubbling gaseous (dry) hydrochloric acid into anhydrous methanol. The concentration of the solution is adjusted by titration of aliquots with 0.1 N sodium hydroxide solution).

The ampoule is cooled in liquid nitrogen or a dry-ice bath, and carefully sealed under vacuum. The ampoule is heated at 110°C for 3 hours and cooled to room temperature. The ampoule is opened and the solvent evaporated to dryness depositing the ester hydrochloride as a solid residue.

Methylene chloride (2 ml) is added to the ester hydrochloride and the solution transferred into a small round-bottomed flask which is protected by a calcium chloride tube. Trifluoroacetic anhydride (2 ml) is cautiously added, and the mixture allowed to reach room temperature and then stirred for 2 hours. The solvent is evaporated to dryness and the residue is redissolved in about 5 ml chloroform.

Coating of Support by Optically Active Stationary Phase

Lauroyl-L-valyl-*tert*-butylamide (500 mg) is placed in a porcelain dish and dissolved in 10 ml chloroform, Chromosorb W (4.5 g, acid washed) is added and the mixture is stirred with a glass rod. The dish is then placed on a water bath (in a hood) and the solvent is evaporated while stirring.

Preparation of an Optically Active GC Column

A stainless steel column (4 m × 0.3 mm) is packed with the optically active support in the usual manner and left overnight at 130°C for conditioning.

Separation of the Enantiomers

A gas chromatograph equipped with a flame ionization detector (FID) is used. Operating conditions: column temperature, 110°C; injector, 150°C; detector, 150°C. The D-enantiomer emerges first and is followed by the L-enantiomer.

REFERENCES

1. Dudley, P. A. and Anderson, R. E., "Separation of Polyunsaturated Fatty Acids by Argentation Thin-Layer Chromatography," *Lipids* **10**:113 (1975).
2. Battaglia, R. and Fröhlich, D., "HPLC Separation of *Cis* and *Trans* Monounsaturated Fatty Acids," *Chromatographia* **13**:428 (1980).
3. Berg, A. and Lam, J., "Separation of Polycyclic Aromatic Hydrocarbons by Thin-Layer Chromatography on Impregnated Layers," *J. Chromatogr.* **16**:157 (1964).
4. Gilon, C., Leshem, R. and Grushka, E., "Structure-Resolution Relationship. I. The Effect of the Alkylamide Side-Chain of Aspartyl Derivatives on the Resolution of Amino Acid Enantiomers," *J. Chromatogr.* **203**:365 (1981).
5. Charles, R., Beitler, U., Feibush, B. and Gil-Av, E., *J. Chromatogr.* **112**:121 (1975).

RECOMMENDED READING

Conacher, H. B. S., "Chromatographic Determination of *Cis-Trans* Monoethylenic Unsaturation in Fats and Oils—A Review, *Journal of Chromatographic Science* **14**:405–411 (1976).
Foster, R., *Organic Charge-Transfer Complexes*, Academic Press, 1969.
Krull, I. S., "The Liquid-Chromatographic Resolution of Enantiomers," in: *Advances in Chromatography* (J. C. Giddings, E. Grushka, J. Cazes and P. R. Brown, Eds.), Vol. 16, p. 175, Marcel Dekker, Inc., 1978.
Morris, L. J. and Nichols, B. W., "Argentation Thin-Layer Chromatography of Lipids," in: *Progress in Thin-Layer Chromatography and Related Methods* (A. Niederwieser and G. Pataki, Eds.), Vol. 1, p. 75, Ann Arbor Science Publishers, 1970.

3
Reaction Chromatography

INTRODUCTION

In recent years, new approaches have been developed whereby chemical reactions can be performed on a gas chromatographic column or, if desirable, the substance undergoes reaction in a suitable pre-column reactor connected to the gas chromatograph. In the latter case, the reaction products obtained are immediately passed into the GC column for separation and collected individually if further identification is required. This technique is called reaction chromatography and is suitable for typical organic reactions such as dehydration, hydrogenation, hydrogenolysis, ozonolysis, etc.

The combination of chemical and chromatographic methods has proved to be a more efficient tool for identification of minute quantities of organic compounds than either of these methods individually. Recently, TLC-GLC combinations have been introduced: the compounds are allowed to react on the thin-layer plate and, after development, the various groups of products are isolated and subjected to GLC analysis. The main advantage of the combination of TLC with GLC stems from the quite different distribution factors which control the separation process. In GLC the separation is principally determined by the volatility of the substance, whereas in TLC the distribution coefficient is mainly determined by the type of the functional groups, volatility playing a less important role.

Functional-group analysis by GC effluent chromatography and the thermomicrochromatographic (TAS) method are also included in this chapter. The first technique involves the separation of the components by GC and their subsequent identification by the performance of characteristic reactions with various reagents.

Although a large number of organic substances can be analyzed directly by GLC, many others, such as carbohydrates, amino acids, etc., contain polar functional groups or are of high molecular weight and therefore are unsuitable for such analysis. In these cases, chemical derivatization, such as silylation, is used to impart to these additives the necessary thermal stability, volatility and desirable chromatographic behavior for GLC analysis.

Chemical derivatization can be carried out prior to, during or after the chromatographic process.

The TAS method is based on the volatilization of organic compounds (on a microscale) from various natural sources and products of micro-reactions performed in the TAS oven and their direct absorption on thin-layer plates for separation and identification. Furthermore, the separated products can be extracted from the thin-layer plates and subjected to GC analysis.

FUNCTIONAL GROUP ANALYSIS BY PRE-COLUMN REACTION GC [1]

In the following experiment pre-column reactions, such as esterification, reduction, oxidation and hydrolysis of organic compounds, are carried out in a microsyringe. Small volumes of the reagent and the organic compound are injected into a microsyringe and allowed to react there for several minutes. The crude reaction mixture is then injected into a gas chromatograph and the products are identified by comparing their retention times with those of known standards. This technique requires minute quantities of compounds and furnishes information on the functional groups and the yields of the products.

Reagents

The reagents used are: sodium borohydride (saturated solution in ethanol) —for reduction of ketones and aldehydes; potassium permanganate (saturated solution in acetone)—for oxidation of alcohols; sodium hydroxide (saturated solution in ethanol)—for hydrolysis of esters; and methanol with 51% boron trifluoride—for esterification of acids.

Procedure

The liquid reagent (0.5 μl) is spread by means of barrel action onto the wall of a 10-μl microsyringe. One microliter of 0.1% ether solution of the sample is then introduced into the syringe and left to react for the appropriate time, depending on the reagent used, as indicated in Table 1. The sample is then injected into a gas chromatograph. Reference compounds are injected and co-injected for comparative purposes. The yields (conversions) are in the range of 50–100%.

Table 1

Compound	Reagent	Column	Temp. (°C)	Retention Time (minutes)	Identity of Product
Acids					
Succinic acid	Methanol/boron trifluoride	LAC 2R-446	175	3.5	Dimethyl succinate
Glutaric acid	"	"	"	4.9	Dimethyl glutarate
Adipic acid	"	"	"	7.0	Dimethyl adipate
Ketones					
2-Decanone	Sodium borohydride	LAC 2R-446	115	17.8	2-Decanol
2-Nonanone	"	"	"	6.9	2-Nonanol
Cyclohexanone	"	"	"	5.7	Cyclohexanol
Aldehydes					
n-Hexaldehyde	Sodium borohydride	LAC 2R-446	114	4.4	1-Hexanol
n-Heptaldehyde	"	"	"	7.2	1-Heptanol
n-Decaldehyde	"	"	"	31.5	1-Decanol
Alcohols					
1-Decanol	Potassium permanganate	LAC 2R-446	115	11.7	n-Decaldehyde
1-Nonanol	"	"	"	6.4	n-Nonaldehyde
1-Octanol	"	"	"	2.8	n-Octaldehyde
Esters					
n-Decyl butyrate	Sodium hydroxide	LAC 2R-446	>130	11.1	1-Decanol
n-Octyl heptanoate	"	"	"	4.6	1-Octanol

Column

A glass column (1.2 m long, 6 mm o.d.) packed with 10% LAC 2R-446 on Chromosorb W is used. Such a column is stable for about 500 injections.

FUNCTIONAL GROUP ANALYSIS BY GC-EFFLUENT CHARACTERIZATION [2]

The identification of components separated by GC cannot be accomplished with reasonable certainty by relying on retention data alone, particularly when the mixture is heterofundtional.

Although it is possible to analyze a complex mixture by finding suitable liquid phases to effect resolution and identification, this procedure is time-consuming and unrewarding. Recently, this problem has been solved by the collection of the eluted peaks and the employment of IR, MS and coupled techniques. Nevertheless, the failure of the GC method to provide complete qualitative analysis is due primarily to the inability of the gas chromatograph to determine or distinguish organic functionality.

In the following method different functional group classification reagents are used, which form visible colored products with the GC-effluents.

Procedure

A conventional, packed, column-type gas chromatograph with a thermal-conductivity detector is used. The functional group classification of the eluted chromatographic peaks is accomplished by means of a stream-splitting device which is attached to the exit tube of the thermal-conductivity cell. The chromatographic effluent is divided into three equal streams, each of which is allowed to bubble through a vial containing an appropriate classification reagent. When a large number of reagents is used, an even and continuous flow in all the vials cannot be obtained and handling becomes rather cumbersome. It is therefore more convenient to run a second sample with another set of reagents.

In the following experiments nine functional groups are considered: alcohols, aldehydes, ketones, esters, unsaturated aliphatic and aromatic hydrocarbons, amines, alkyl halides, sulfur compounds and nitriles. The reagent solutions used for characterization of these classes of compounds are as follows:

For alcohols: (a) nitrochromic acid (0.5 ml of 7.5 N HNO$_3$ plus 1 drop of 1% potassium dichromate)—turns from bright yellow to blue-gray—used for primary and secondary alcohols; (b) ceric nitrate (5 drops of reagent plus 5 drops of water)—turns from yellow to amber—used for all aliphatic alcohols.

For aldehydes and ketones: 2,4-dinitrophenylhydrazine solution (10 drops) —forms yellow or orange precipitate.

For Esters: ferric hydroxamate test: 10 drops of 1 N hydroxylamine hydrochloride in methanol plus 3 to 4 drops of 2 N alcoholic potassium hydroxide or until solution turns blue. After adding sample to the solution 5 to 6 drops of 2 N HCl are added until solution is clear and colorless. Upon addition of 1 to 2 drops of 10% FeCl$_3$, it becomes colorless or red.

For alkyl halides: (a) alcoholic AgNO$_3$ (10 drops)—forms white precipitate; (b) mercurous nitrate (10 drops of 7.5 N HNO$_3$ plus 5% mercurous nitrate solution)—forms yellow to orange precipitate with iodides, white precipitate with chlorides, and white or gray precipitate with bromides.

For amines: (a) benzenesulfonyl chloride (Hinsberg test): 5 drops of pyridine, 1 drop of 5% sodium hydroxide. After addition of the sample, 1 to 2 drops of benzenesulfonyl chloride are added. Colorless to yellow for primary or secondary aliphatic amines. Tertiary amines give rise to a deep purple color. (b) Sodium nitroprusside (Rimini and Simon test for primary and secondary amines): 10 drops of water, 2 drops of acetone and 1 drop of 1% sodium nitroprusside. Primary amine gives red color. Upon addition of 1 to 2 drops of acetaldehyde, secondary amine turns blue. Omitting acetone permits test for secondary amine directly.

For alkyl nitriles: ferric hydroxamate: 10 drops of 1 N hydroxylamine hydrochloride in propylene glycol plus 2 drops of 1 N potassium hydroxide in propylene glycol. After sample passes through the reagent it is heated to boiling and cooled. Solution becomes clear and colorless. On addition of 1 to 2 drops of 10% ferric chloride, it turns a red-wine color.

For mercaptans: (a) alcoholic silver nitrate (10 drops, 2%)—forms white precipitate (H$_2$S forms black precipitate); (b) lead acetate (10 drops, saturated alcoholic)—forms yellow precipitate (H$_2$S forms black precipitate); (c) isatin (10 drops, 1% in concentrated sulfuric acid)—gives green color; (d) sodium nitroprusside (10 drops of 95% ethyl alcohol plus 2 drops of a 5% KCN plus 1% sodium hydroxide solution). Two to three minutes after sample addition, 5 drops of 1% sodium nitroprusside solution are added, giving rise to a red color.

For alkyl sulfides: sodium nitroprusside (same as test (d) for mercaptans).

For aromatic nuclei and unsaturated aliphatics: formaldehyde-sulfuric acid

(Le Rosen test): 10 drops of concentrated sulfuric acid plus 1 drop of 37% formaldehyde give a wine color.

The use of the above-mentioned reagents is obviously dependent on the types of compounds expected and the desired level of response of the reagent. For example, the potassium dichromate-nitric acid (nitrochromic acid) test was found most suitable for alcohols, except tertiary alcohols for which the ceric nitrate reagent is preferred.

In some cases, the response of two or more reagents must be considered in combination. For example, since both aldehydes and ketones give a positive reaction with 2,4-dinitrophenylhydrazine, Schiff's reagent is used to distinguish the former from the latter.

DETERMINATION OF EPOXIDE POSITION AND CONFIGURATION BY COMBINATION REACTION TLC/GC [3]

Epoxides are abundant in many natural and synthetic products. The following procedure enables one to locate the position of epoxide groups in microgram amounts of a compound by reacting it in a halogenated solvent with dry, powdered HIO_4 for 5 minutes. The aldehyde and ketone fragments produced by cleavage between the carbon atoms of the epoxide group are determined by GC of the reaction solution. Epoxides are detected by applying them over a phosphoric acid spot on a silica-gel TLC-plate and allowing the reaction to proceed for 1 hour before plate development. The products, being more polar than the original compound, remain near the origin, and the original spot is no longer visible. The geometrical configuration of disubstituted epoxides is determined as above, except that allowance is made for a 5-minute reaction interval prior to plate development.

Determination of Epoxide Position

Aldehydes are obtained from compounds in which the carbon atoms linked to the epoxide oxygen have a hydrogen substituent, and ketones are obtained when these carbon atoms are dialkyl substituted.

$$R'-\underset{\underset{O}{\diagup}}{\overset{\overset{R}{|}}{C}}-CH-R'' \xrightarrow{HIO_4} R'-\overset{\overset{R}{|}}{C}=O + R''CHO$$

$$R, R', R'' = alkyl$$

Configuration of Epoxides

The rapid reaction of unhindered disubstituted *cis*-epoxides with phosphoric acid and the relatively slow reaction of the corresponding *trans*-epoxides form the basis of the procedure advanced as a means of differentiating these *cis*- and *trans*-epoxides. In the 5-minute reaction period chosen for this determination, *cis*-epoxides, 1,2-epoxides, and trisubstituted epoxides are completely or almost completely subtracted. *Trans*-epoxides or hindered epoxides are subtracted partially or in trace amounts.

Equipment and Materials

A dual-column flame-ionization gas chromatograph fitted with 6 mm o.d. glass or stainless steel columns is used, helium serving as the carrier gas. One column is 1.2 m long and contains 5% diethylene glycol succinate (DEGS) on 60- to 80-mesh base-washed Chromosorb W; the other is 2.75 m long and contains 20% DEGS on 60- to 80-mesh acid-washed Chromosorb W. The temperature of the injection port and detector is 220°C. The column temperature is programmed, being selected to permit the determination of both the original compound and the reaction products.

If the periodic acid is supplied as H_5IO_6 it is ground to a powder in a mortar and dried to constant weight in an evacuated drying pistol over P_2O_5 at 78°C (two moles of water are lost to give HIO_4).

Procedure for Determining the Position of the Epoxide Group

To the samples (1–100 μg) in 100 μl of solvent in a screw-cap bottle, dry HIO_4 (in an amount 340 times the sample weight plus 2 mg extra) is added. The mixture is shaken for 5 minutes and centrifuged, if necessary. Methylene chloride is used to extract the products which are injected into the 1.2-m, 5% DEGS column to observe fragments with retention times equal to or greater than that of heptanal. Tetrachloroethane is similarly used with the 2.75-m, 20% DEGS column for observation of fragments with lower retention times.

Procedure for Subtracting cis- and trans-Epoxides by Reaction TLC

The silica-gel plates are prewashed with ether. Five microliters of aqueous phosphoric acid (10%) are placed on each spot to be used for subtraction of epoxides, leaving a blank space alongside for determination of the R_f value of the untreated sample. After a few minutes, when the acid appears dry, another 5 μl of 10% phosphoric acid are added on the original spot. After allowing the acid to dry for 20 minutes, 20–100 μg of the compound are spotted both on the phosphoric acid and on the blank space alongside. The plate is set aside for 1 hour, and then developed twice with 20% by volume of ether in hexane. Compounds are made visible with iodine vapor and/or by spraying with an aqueous solution containing 8% by volume of 85% H_3PO_4 and 2% cupric acetate, followed by heating in an oven at 130°C for 15 minutes.

Procedure for Determining cis-*Epoxide Configurations by TLC*

Silica-gel plates are prewashed with ether and the phosphoric acid and samples are applied as described above. The treated samples are allowed to react with phosphoric acid for 5 minutes and developed immediately with ether-hexane mixture. The compounds are detected by iodine vapors or by spraying with cupric acetate in phosphoric acid reagent as described above.

REACTION GC IN SEALED GLASS CAPILLARIES [4]

In the technique of reaction GC, microgram or sub-microgram quantities of organic material are condensed from the GC-effluent in glass capillary tubing. The tubing is then filled with a catalyst and a reactive gas, such as hydrogen, oxygen or ozone, and sealed.

Reaction is effected by heating the sealed capillary, after which all the products of the reaction are re-introduced into the gas chromatograph by crushing the capillary within the injection port.

Procedure

The yield of reaction products obtained from a few micrograms of material is usually sufficient for GC and MS analysis.

The collection of purified material from the outlet of the gas chromatograph is illustrated in Fig. 1. The glass capillary G (1.45 mm o.d. and 115 mm long) is introduced into the connector C. Since it is essential to confine the condensed material to the center of the capillary tube, the heated outlet line A is extended by means of an aluminum heat conductor D or by wrapping aluminum foil or asbestos tape around the capillary. If a catalyst E is later required, it is spread over 3 or 4 mm on the wall of the capillary tube. Solid

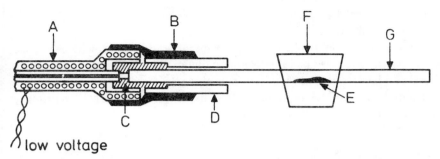

low voltage

FIG. 1. Collection of the sample: A—heated GC outlet line; B—aluminum foil and asbestos tape; C—PTFE connector; D—aluminum heat conductor; E—catalyst; F—solid carbon-dioxide cooling boat; G—glass capillary.

carbon dioxide, contained in a plastic boat F, assists condensation of material from the GC-effluent. When collection is completed, the capillary is removed, closed with tight-fitting caps and stored under solid carbon dioxide until required.

The gases (H_2, N_2, O_3) for filling the CO_2-cooled glass capillaries are delivered at reduced pressure (1–2 psig) for 1 minute. The glass tubing is then sealed with a micro-flame, firstly at the atmospheric end and then at the supply end. Excessive softening of the glass must be avoided in order to prevent the pressurized gas from bursting through the seal.

The reaction conditions are ranged from 30 minutes at −75°C for ozonolysis to 5 minutes at 300°C for hydrogenolysis. After the reaction, the external surface of the sealed capillary is washed and dried in preparation for re-injection of the reaction products.

The reaction products may be injected with any device capable of breaking the capillary (after it has reached the port temperature) within the injection port.

The dual-column gas chromatograph is equipped with two 6 m × 2 mm glass columns, one packed with 3% SF-96 on Chromosorb G and the other with 3% Carbowax 20M on Chromosorb G.

Micro-Dehydration

Each of the *n*-C_5 to -C_{10} alcohols is collected in turn from the GC outlet, on a small bed of acidic alumina in a capillary. The capillary is then filled with nitrogen, sealed, and heated to 300°C for 3 minutes. The reaction products (mainly olefins) are re-injected into the GC system using the glass capillary crushing device. For each alcohol at temperatures below 300°C a mixture of olefins is obtained.

Basic alumina yields olefins with a terminal double bond, whereas acidic alumina yields all five possible isomeric olefins. The position of the double bond can be determined by micro-ozonolysis.

Secondary alcohols can generally be dehydrated more easily than primary alcohols, and at lower temperatures (250°C).

Micro-Reduction of Aldehydes and Ketones to Alcohols

The reduction of carbonyl groups without the loss of unsaturation is a desirable complement to hydrogenation. A small quantity of sodium borohydride (1 mg) is used to reduce aldehydes, such as n-heptanal, and cis-hept-4-enal, to primary alcohols at 160°C within 3 minutes. The ketone, octan-4-one, is reduced more slowly at 160°C to yield the secondary alcohol.

The usual addition of water as a second rection step, to liberate the alcohol from the sodium complex, is not necessary when using the above-mentioned carbonyls. In other cases, it is essential to open the capillary after the reaction with sodium borohydride, to add 1–2 µl of water, reseal the capillary and heat it prior to re-injection of the products into the gas chromatograph.

Micro-Oxidation of Alcohols to Aldehydes and Ketones

The following alcohols can be used as test compounds for the oxidation reaction: n-butanol, n-hexanol, n-heptanol, hexan-2-ol, hexan-3-ol, heptan-2-ol, 2- and 3-methylbutanols. Finely ground potassium dichromate crystals (0.25–0.5 mg) are used to oxidize the alcohols at 250–280°C for 3 minutes. The optimum reaction conditions vary for primary alcohols which require more careful control than the secondary alcohols.

SUBTRACTION GC [5]

Gas chromatography has the inherent capability of detecting very small amounts of a compound even in complex mixtures. However, it is usually inadequate as a diagnostic tool for determining the chemical structure of a compound in a completely unknown mixture.

Reaction GC techniques, such as pyrolysis, hydrogenation, hydrogenolysis, dehydrogenation and carbon-skeleton chromatography, have proved useful for this purpose. The subtraction-loop technique furnishes information on functional groups of a compound by forming nonvolatile products with particular chemicals. Abstraction of a compound from a mixture usually takes place on-stream, the active reagent being distributed on a convenient solid support material packed in a metal loop.

A subtraction loop is fabricated from a 6-inch length of $\frac{1}{4}$-inch stainless steel tubing coiled once with the ends facing in opposite directions, as shown in Fig. 2. The loop holds about 800 mg of support material. The loop can be mounted either between the injection post and the column or between the column and the detector. The packings for the subtraction loops are prepared as follows: for carboxylic acids—1 part of zinc oxide is physically mixed with 10 parts by weight of the LAC-2R-446–phosphoric-acid packing; for alcohols—1 part of powdered boric acid is similarly mixed with 20 parts by weight of Carbowax 20M. The FFAP and aromatic amines used to subtract aldehydes, ketones and epoxides are deposited on the support by evaporation from chloroform solution to give the following compositions: 5% o-dianisidine on 70- to 80-mesh Anakrom ABS; 20% benzidine on 60- to 80-mesh acid-washed Chromosorb P; and 20% FFAP on 60- to 80-mesh acid-washed Chromosorb P. When using the amine reagents, the last half inch of the 6-inch loop is packed with uncoated Anakrom ABS to minimize substrate bleed. For epoxides the loop is filled with 100 to 200 mg Chromosorb W coated by 5% by weight concentrated phosphoric acid, followed by the coated support (as in the analytical column) to complete the filling. The phosphoric acid is evaporated in the support from a water solution. Packings are held in place with glass-wool plugs.

In general, it is necessary to inject two samples, one of them to be run without an abstractor (loop), so that one knows which peaks are removed by the chemical reaction loop and cannot be conveyed to the column by the carrier gas.

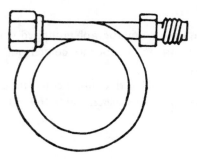

FIG 2. Six-inch subtraction loop.

Table 2

Abstractor	Materials Removed
Molecular sieves, 5Å	n-Alkanes
Mercuric acetate-ethylene glycol	Alkenes
Maleic anhydride-silica gel	Dienes
Boric acid	Alcohols
Sodium bisulfite-ethylene glycol	Aldehydes
Benzidine	Aldehydes, Ketones
o-Dianisidine	Aldehydes
Phosphoric acid	Epoxides
Sulfuric acid	Aromatics, Alkenes, Alkynes
Zinc oxide	Acids

Reaction subtraction loops expose a large surface area of reagent to the sample, thus ensuring complete abstraction with normal flow rates of carrier gas, provided that the reaction is rapid.

Table 2 lists the selective abstractors and the materials removed by them.

The following experiments illustrate the utility of zinc oxide and boric acid as subtraction agents. In addition, two aromatic amines and phosphoric acid are shown to be useful reagents: o-dianisidine (3,3'-dimethoxybenzidine) quantitatively subtracts aldehydes; benzidine removes most ketones and aldehydes; and phosphoric acid subtracts epoxides. Each reagent is diluted with or coated on the packing material or support. The reaction loop is connected to the analytical column, and gas chromatographic determinations are made with and without the loop to ascertain the degree of subtraction.

Equipment and Materials

A flame-ionization gas chromatograph equipped with a 10-foot, $\frac{1}{4}$-inch o.d. glass column filled with 5% Carbowax coated on Anakrom (70–80 mesh) is used.

For acids a 4-foot, $\frac{1}{4}$-inch o.d. column containing 25% LAC-2R-446 plus 2% phosphoric acid on acid-washed Celite (60–80 mesh) is employed.

Procedure

The compounds are injected into the gas chromatograph as hexane or methylene chloride solutions in 2% w/v concentrations. The amount of compound chromatographed varies from less than 1 μg to 1 mg; 60 μg is the customary sample size.

The analytical columns and subtraction loops are generally operated isothermally in the range of 75°–200°C. Nitrogen, at a flow rate of 40 ml/minute, is used as the carrier gas.

OZONOLYSIS-GC OF ORGANIC COMPOUNDS IN THE MICROGRAM RANGE [6]

Determination of the position of carbon-carbon double bonds by means of ozonolysis is common practice in conventional organic chemistry. With the aid of GC the technique can be extended to quantities of sample very much smaller than those normally used.

The position of unsaturation of ozonides can be determined by oxidative, reductive or pyrolytic cleavage techniques. Thermal decomposition of ozonides leads to the formation of acidic and aldehydic fragments. Pyrolysis of ozonides in the presence of a hydrogenation catalyst (Pd on C) yields only aldehydic products. Similar results are obtained by hydrolysis in the presence of zinc or by reduction with triphenyl phosphine:

$$
\begin{array}{c}
\text{O—O} \\
\overset{\diagup \qquad \diagdown}{RHC \qquad\qquad CHR'} + Ph_3P \longrightarrow RCHO + R'CHO + Ph_3PO \\
\underset{\diagdown \quad \diagup}{O}
\end{array}
$$

The latter method is used in the following experiment which requires 1 to 5 μg of sample.

Equipment and Materials

The ozonizing unit is shown in Fig. 3. The oxygen is passed through a Linde 5Å molecular sieve prior to entering the ozonator.

The carbon disulfide and pentyl acetate used should be chromatoquality reagents. The indicating solution used to reveal an excess of ozone consists of 5% potassium iodide in 5% aqueous sulfuric acid with starch added.

Procedure

Solutions of the ozonolysis products are analyzed on a gas chromatograph equipped with a flame-ionization detector and a 12-foot, $\frac{1}{4}$-inch o.d. glass

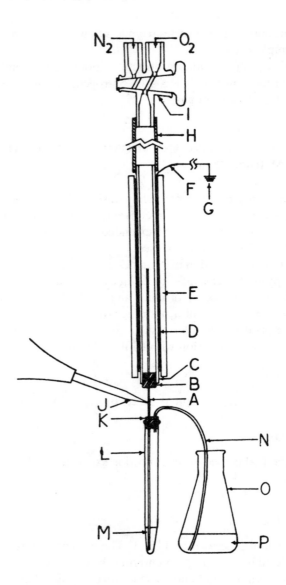

FIG. 3. Setup for ozonization of compounds. A through J are parts of micro-ozonizer: A—needle stock tubing; B—rubber stopper (injection-port septum); C—glass tubing; D—aluminum foil; E—rubber tubing; F—copper wire; G—ground; H—rubber tubing; I—three-way stopcock; J—high-voltage source (vacuum tester). K—rubber stopper (injection-port septum); L—reaction tube; M—solution of compound; N—Teflon tubing; O—10-ml Erlenmeyer flask; P—indicating solution (4 ml). Cold bath held at −70°C is not shown.

column containing 5% Carbowax 20M on 60- to 80-mesh Gas Chrom P. The flow rate of nitrogen carrier is 30 to 60 ml/minute. The programming is 6–7°C/minute up to 200°C.

Twenty-five micrograms of the pure compound in 100 μl of methylene chloride or pentyl acetate are introduced into tube L (Fig. 3). Oxygen (10 ml/minute) is passed into the solution while the tube and solution are cooled to ca. −70°C in an acetone-dry ice bath. The energized vacuum

Table 3

Compound	Products Identified
⬡—CH=CH—C(=O)—CH$_3$	⬡—CHO
⬡—CH=CH$_2$	⬡—CHO
CH$_2$=CHOCH$_2$CH$_3$	H—C(=O)—OCH$_2$CH$_3$
⬡—C(CH$_3$)=CH$_2$	⬡—C(=O)—CH$_3$
CH$_3$CH=CH(CH$_2$)$_4$CH$_3$	CH$_3$CHO, CH$_3$(CH$_2$)$_4$CHO
CH$_3$O—⬡—CH=CHCH$_3$	CH$_3$CHO, CH$_3$O—⬡—CHO
(CH$_3$)$_2$C=CH—(cyclopropane ring with CH$_3$ and COOCH$_3$)	CH$_3$—C(=O)—CH$_3$
CH$_3$(CH$_2$)$_7$CH=CH(CH$_2$)$_7$COOCH$_3$	CH$_3$(CH$_2$)$_7$CHO
CH$_3$(CH$_2$)$_4$CH=CHCH$_2$CH=CH(CH$_2$)$_7$COOCH$_3$	CH$_3$(CH$_2$)$_4$CHO
(thiophene)—CH=CHCOOCH$_3$	(thiophene)—CHO

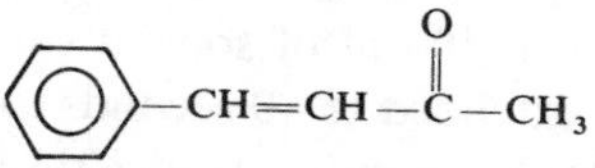

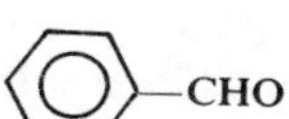

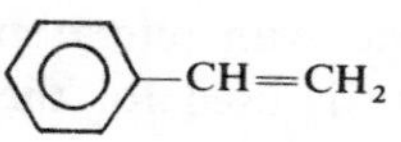

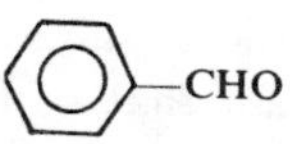

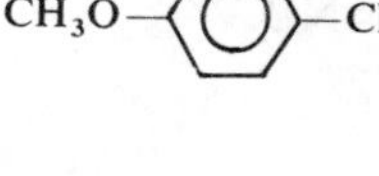

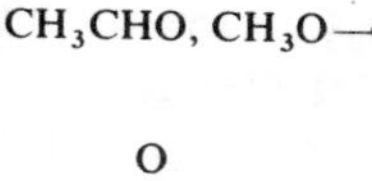

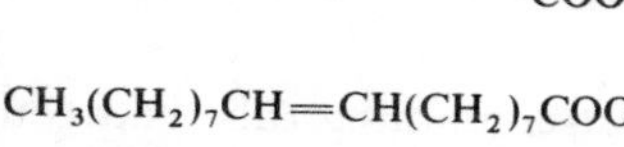

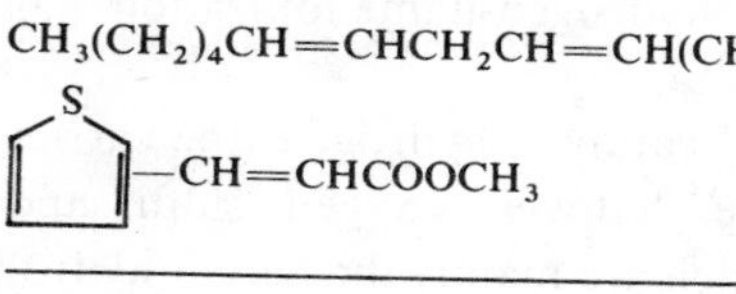

tester or Tesla coil J (source of high voltage) is applied to electrode A to generate ozone. When the blue color of excess ozone becomes visible in the indicating solution, the vacuum tester is removed. Ozone generation usually requires 10 to 15 seconds. The Teflon stopcock I is turned in order to purge the solution with nitrogen for 15 seconds, thus replacing the oxygen. The cold bath is removed, tube L is slipped off rubber stopper K, and about 1 mg of powdered triphenyl phosphine is dropped onto the solution. The tube is immediately stoppered and swirled to dissolve the powder. When the solution reaches room temperature, a 20-μl aliquot is injected into the gas chromatograph for analysis. Identifications are based on the retention times of known compounds. As little as 1 μg of compound in 100 μl of solvent can be analyzed.

Samples (200–400 μg) of tung and linseed oils in 100 μl of solvent can also be analyzed. The ozonization period is then extended to 90 seconds to allow for possible slow reaction of conjugated double bonds. Ozonization in carbon disulfide is suitable for the analysis of products with retention times greater than that of butyraldehyde; pentyl acetate is used for the analysis of the smaller fragments.

Table 3 summarizes the products obtained by micro-ozonolysis of various compounds (5 μg of compound per analysis).

This method provides a simple means for analyzing compounds of low molecular weight, since no evaporations are required and recoveries at the C_3 or higher levels are good. When sufficient material is available, thermal conductivity detection may be employed, especially if product trapping is desirable.

DETERMINATION OF THE CARBON SKELETON AND OTHER STRUCTURAL FEATURES OF ORGANIC COMPOUNDS BY GC [7]

Chemists, especially those dealing with natural products, are frequently confronted with the problem of identifying substances which they have isolated in insufficient amounts.

The following technique utilizes a hot catalyst-containing tube between the injection port and the column of a hydrogen-flame-ionization gas chromatograph.

On sweeping the sample over the heated catalyst, hydrogen (the carrier gas) saturates multiple bonds and replaces halogen, oxygen, sulfur and nitrogen atoms. The resulting products, which are mainly the parent hydrocarbon and/or the next lower homolog, are identified by their elution time through an appropriate column. These data aid one in determining the carbon skeleton of the compound and often the position of functional groups. The method is rapid and applicable for the determination of the structure of organic compounds containing as many as nine connected carbon atoms.

Equipment and Materials

Catalyst assembly

The catalyst assembly is shown in Fig. 4. It consists of a $9\frac{1}{2}$ inch-long aluminum tube D ($\frac{1}{2}$-inch o.d., $\frac{1}{8}$- to $\frac{3}{16}$-inch i.d.) screwed onto the usual injection port of the gas chromatograph by means of fitting F. The catalyst charge is placed in tube D. A compression "T" fitting holds the injection-port assembly C and the hydrogen-gas inlet B. Lead washers are used to make the fitting F and the injection port gas-tight. The temperature of the catalyst tube, determined by a thermocouple embedded in the tube $1\frac{1}{2}$ inches from the end holding fitting F, is regulated by a heating jacket E, constructed by wrapping nickel-chromium resistance wire around asbestos tape on a glass tube; the ends of the wire are connected to a variable transformer.

An appropriate length of magnesia pipe insulation A, coated with an asbestos cement to prevent shredding, covers the heating jacket.

Flame-ionization gas chromatograph

Hydrogen (the carrier gas) is introduced via the catalyst tube into the column. The hydrogen inlet of the usual commercial detector is closed because nitrogen is unnecessary.

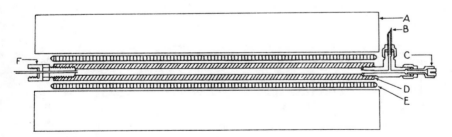

FIG. 4. Catalyst assembly: A—insulation; B—hydrogen inlet; C—injection port; D—aluminum tube; E—heating jacket; F—fitting that attaches catalyst assembly to gas chromatograph injection port.

Table 4

Type	Structure	Principal Products	Catalyst; Temp. (°C)
Anhydride	$[(CH_3)_2CHCO]_2O$	C_3	1% Pd-GCP*; 260
Acid	CH_3COOH	C_1	1% Pd-PG**; 265
	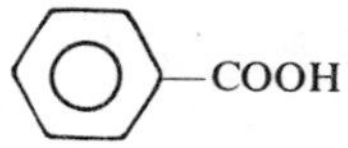 (benzene ring)—COOH	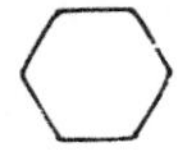 (cyclohexane ring)	1% Pd-GCP; 250
Alcohol	CH_3OH	C_1	1% Pd-GCP; 265
	$CH_3CH_2CH_2OH$	C_3	1% Pd-GCP; 265
	$CH_3(CH_2)_4OH$	C_4, C_5	5% Pt-PG; 210
	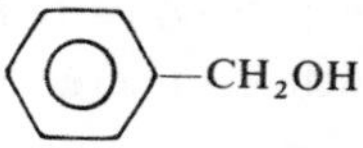 (benzene ring)—CH_2OH	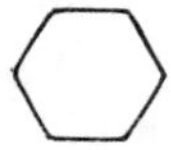 (cyclohexane ring)	1% Pd-GCP; 250
	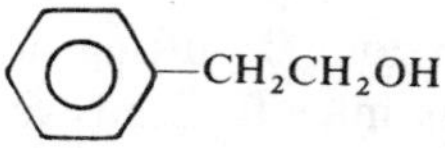 (benzene ring)—CH_2CH_2OH	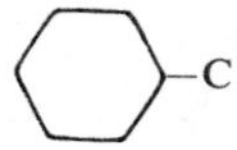 (cyclohexane ring)—C	1% Pd-GCP; 250
Aldehyde	$CH_3CH_2CH_2CHO$	C_3	1% Pd-PG; 265
Amide	$CH_3CH_2CONH_2$	C_3	5% Pt-PG; 210
	$C_3H_7CONH_2$	C_3, C_4	5% Pt-PG; 210
Amine	$C_2H_5NH_2$	C_2	5% Pt-PG; 210
Ester	$C_3H_7OOC \cdot CH_2CH_3$	C_3, C_2	1% Pd-PG; 250
	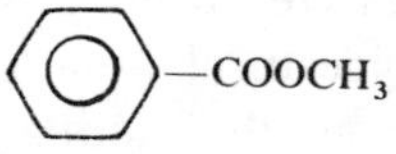 (benzene ring)—$COOCH_3$	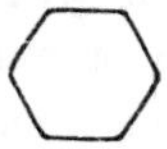 (cyclohexane ring)	1% Pd-GCP; 250
Ether	$C_2H_5OC_2H_5$	C_2	1% Pd-PG; 260
	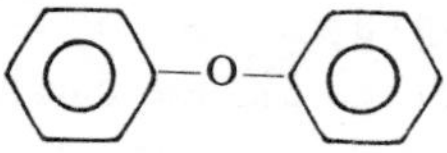 (benzene ring)—O—(benzene ring)	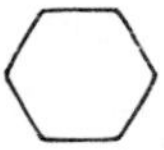 (cyclohexane ring)	1% Pd-GCP; 260
Halide	CH_2Cl_2	C_1	1% Pd-PG; 187
	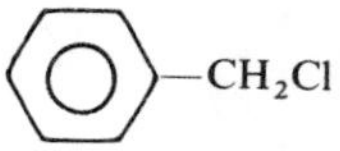 (benzene ring)—CH_2Cl	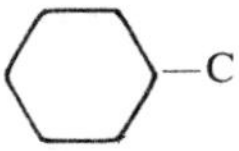 (cyclohexane ring)—C	1% Pd-GCP; 250
Ketone	$(CH_3)_2CO$	C_3	5% Pt-PG; 210
Phenol	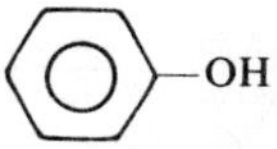 (benzene ring)—OH	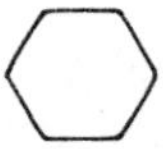 (cyclohexane ring)	1% Pd-PG; 260
Sulfide	CH_3CH_2SH	C_2	1% Pd-PG; 260
Epoxide	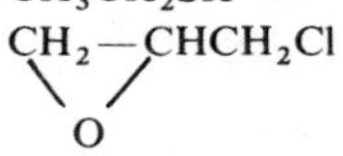 $CH_2\!\!-\!\!CHCH_2Cl$ (with O bridge)	C_3	5% Pt-PG; 210

* GCP = Gas Chrom P.

** PG = porous glass.

Two gas chromatograph columns are used: an 8-foot, $\frac{1}{8}$-inch o.d. stainless steel column containing 5% silicone gum SE-30 on 60- to 80-mesh acid-washed Chromosorb W; and a 15-foot, $\frac{3}{16}$-inch o.d. copper column containing 5% squalane on 60- to 80-mesh acid-washed Chromosorb W.

Catalysts

Palladium (1%) or platinum (5%) on 60- to 80-mesh porous glass and palladium (1%) on Gas Chrom P are used.

Procedure

The catalyst tube is maintained at a definite temperature (usually between 125° and 290°C) by adjusting the variable transformer. The chromatographic column is maintained at 25°C and the hydrogen flow at 20 ml/minute. About 0.02 μl (20 μg) of pure compound is introduced into the injection port with a microsyringe.

Hydrocarbon mixtures of known composition are injected periodically to establish their retention times. The retention times of "unknowns" are then related to those of the "knowns" in order to establish identities.

When hydrogen is injected into the catalyst chamber, it reacts with the reactive groups of the compounds. Inorganic products, such as water, hydrogen chloride and ammonia, do not produce any signal with the flame-ionization detector. Paraffin hydrocarbons pass through the chamber and are eluted as sharp peaks, whereas polar compounds, e.g., ketones and carboxylic acids, are detained by the catalyst and their elution patterns tend to drag or tail.

Table 4 lists the hydrocarbon products obtained from various types of compounds.

A THERMOMICRO-PROCEDURE FOR RAPID EXTRACTION AND DIRECT APPLICATION IN TLC (TAS METHOD) [8]

Many organic and some inorganic substances are volatile at high temperatures. Therefore, if these substances are present in mixtures with non-volatile materials, they can be separated by application of heat, either by distillation or sublimation. The TAS method (*t*hermomicro-*a*pplication of *s*ubstance) is a procedure for the isolation and separation of many substances from solid materials and their direct transfer to the starting line on a TLC-plate. The sample is introduced into a glass cartridge with a conical top and heated rapidly for a short time at a pre-set temperature. The emerging vapors are deposited as a spot on the TLC-plate which is then chromatographed in the usual way. Such a spot can also be scraped off the TLC-plate, extracted with a suitable organic solvent and analyzed by GLC.

This technique may be used in the fields of phytochemistry, pharmacognosy, drug research, analysis of food additives and geochemistry, and in synthetic organic and inorganic chemistry.

General Description of the TAS Procedure

The apparatus is shown in Fig. 5. A small amount of sample D is placed in a special glass cartridge B, the end of which can be closed in a simple and effective way by seal A. The charged cartridge is pushed into a metal block furnace C, adjusted to a given temperature. The top of the glass tube projects from the furnace and is pointed toward the starting line of the chromatographic layer F. The TLC-plate is positioned 1 mm from the top and can easily be moved. The volatile materials are thus transferred directly from the sample onto the layer in the form of spots. The applied materials are then chromatographed in the usual way.

Equipment for TAS Procedure

The following three features of the apparatus are of importance: the form of the glass tube (cartridge) and the sealing of the inlet end; provision of a

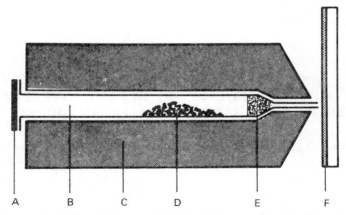

FIG. 5. Cross section of the major components of the TAS oven: A—seal (silicone membrane); B—glass cartridge; C—heating block; D—sample; E—glass wool; F—thin-layer chromatographic layer.

micro-furnace that can be controlled over the temperature range of 50° to 450°C; suitable positioning of the TLC-plate in front of the exit end of the glass tube.

Sample tube (TAS cartridge)

The tube can easily be made from thin-walled, high-melting glass tubing. The tube must fit exactly into the furnace to ensure good heat transfer. The inner diameter of the tip (drawn into a capillary) should be 0.8 to 1 mm. The length of the tube is so chosen that the tip projects only 0.5 to 1 mm from the furnace. The other end, through which the sample is introduced, has a rim; it serves as a seat for the seal and simultaneously as a rest for the sealing clamp. Sealing is achieved by means of a silicone gum membrane, held in place with the clamp. The dimensions of the tube can be reduced or increased. The glass tubes may be discarded after use to save cleaning.

Micro-furnace (TAS oven)

The best heat source for rapid and uniform heating of the tube is an aluminum block heated by a suitable bunsen burner with the necessary controls. The block, with a cylindrical base to accommodate the glass cartridge, has a conical outer shape in order to reduce heat radiation onto the TLC-plate. The temperature is measured with a glass thermometer placed in the block.

Direct electric heating is, however, much more advantageous as the required temperature can be obtained by adjusting a simple bi-metal control. A small thermocouple placed in the block is used for temperature measurement.

Positioning of the TLC-plate

The TLC-plate must be positioned in such a way that it can easily be moved along the starting line, i.e., along a line 15 mm above the lower edge of the plate. The distance between the cartridge tip and the thin chromatographic layer should be only 1 mm and must remain so when the plate is moved.

Some Applications of the TAS Procedure

Analytical phytochemistry and pharmacognosy

Many interesting constituents of plants containing drugs and essential oils can readily be analyzed by the TAS procedure. Generally, 5 to 25 mg of the dry plant material are introduced into the cartridge, heated to 250°C for 90 seconds, and the volatiles are adsorbed directly on the TLC-plate.

In addition, the solid material of plants or pharmaceutical drugs, e.g., caffeine, can be sublimed directly onto a glass slide, and the crystal formed examined with a microscope. The pyrolysis product of opium can also be identified by this method.

Food additives

It is possible to sublime biphenyl directly from a piece of lemon peel or from citrus wrapping papers onto the TLC-plate without solvent extraction and then to identify it. Other preservatives can also be determined, e.g., benzoic acid, salicylic acid, boric acid and p-hydroxybenzoic acids.

Organic chemistry

Distillation and sublimation are simple, frequently applied purifying operations. It is possible to sublime the substance in the form of a uniform layer directly onto a potassium-bromide disc, and to measure its IR spectrum.

In the structural elucidation of organic compounds, methods such as zinc-powder distillation, alkali fusion, sulfur and selenium dehydrogenation can be used on microscale. The use of selenium and sulfur is described in the following experiments:

Dehydrogenation of cholesterol: A mixture of cholesterol (200–300 μg) and selenium (50–100 mg) is placed in a tube fitted for the TAS oven. Heating at 350–400°C yields a complex mixture of products. The main product on the thin-layer plate is chrysene.

Dehydrogenation of abietic acid: A mixture of abietic acid (50 μg) and sulfur (10 mg) is heated in the TAS oven at 230–280°C. The main product is retene.

Organic Geochemistry

The TAS procedure can be employed for samples of oil shale, coal, peat and bitumen materials, such as hydrocarbons, pigments, acids, sterols, etc.

REFERENCES

1. Fredricks, R. M. and Taylor, R., "Functional Group Analysis in Gas Chromatography," *Anal. Chem.* **38**:1961 (1966).
2. Walsh, J. T. and Merritt, C., "Qualitative Functional Group Analysis of Gas Chromatographic Effluents," *Anal. Chem.* **32**:1378 (1960).

3. Bierl, B. A., Beroza, M. and Aldridge, M. H., "Determination of Epoxide Position and Configuration at the Microgram Level and Recognition of Epoxides by Reaction Thin-Layer Chromatography," *Anal. Chem.* **43**:636 (1971).
4. Stanley, G. and Kennett, B. H., "Reaction Gas Chromatography of Microgram and Sub-Microgram Samples Using Sealed Glass Capillaries," *J. Chromatogr.* **75**:304 (1973).
5. Bierl, B. A., Beroza, M. and Ashton, W. T., "Reaction Loops for Reaction Gas Chromatography, Subtraction of Alcohols, Aldehydes, Ketones, Epoxides and Acids and Carbon Skeleton Chromatography of Polar Compounds," *Mikrochim. Acta*, 637 (1969).
6. Beroza, M. and Bierl, B. B., "Rapid Determination of Olefin Position in Organic Compounds in Microgram Range by Ozonolysis and Gas Chromatography," *Anal. Chem.* **39**:1131 (1967).
7. Beroza, M., "Determination of the Chemical Structure of Microgram Amounts of Organic Compounds by Gas Chromatography," *Anal. Chem.* **34**:1801 (1962).
8. Stahl, E., "Thermomicroprocedure for Rapid Extraction and Direct Application in Thin-Layer Chromatography," *Analyst*, 723 (1969).

RECOMMENDED READING

Berezkin, V. G., *Analytical Reactions Gas Chromatography*, Plenum Press, 1968.
Beroza, M., "Determination of the Chemical Structure of Organic Compounds at the Microgram Level by Gas Chromatography," *Accounts Chemical Research* **3**:33–38 (1970).
Brown, R. F. C., *Pyrolytic Methods in Organic Chemistry*, Academic Press, 1980.
"Derivatization in Chromatography" (a series of papers), *Journal of Chromatographic Sciences* **17** (1979).
Grob, R. L., *Modern Practice of Gas Chromatography*, Wiley-Interscience, 1977.
Ikan, R., "Reaction and Combination Chromatography," *Reviews in Analytical Chemistry*, pp. 175–192, Freund Publishing House, Israel, 1972.
Lawrence, J. P. and Frei, R. W. (Eds.), *Chemical Derivatization in Liquid Chromatography*, Journal of Chromatography Library No. 7, Elsevier, 1976.
Leathard, D. A. and Shurlock, B. C., *Identification Techniques in Gas Chromatography*, Wiley-Interscience, 1970.

4
Determination of Food Constituents by Chromatographic Techniques

INTRODUCTION

The food-processing industry requires analytical information about ingredients and products in order to maintain quality and provide nutrition-value information for regulatory compliance. Recent developments in chromatography have provided new and reliable methods for the separation and quantitative characterization of minute amounts of essential dietary components. They also facilitate the detection of food toxicants and possible food adulterants.

HPLC DETERMINATION OF SUGARS IN FOOD PRODUCTS [1]

In view of the increased concern about nutrition and food labelling, reliable quality control is needed for a wide variety of foods and food ingredients, particularly in regard to their sugar content. Several colorimetric and GC methods have been developed. However, the colorimetric methods are non-specific and lengthy, and the trimethylsilyl (TMS) derivatives for GC are extremely sensitive to moisture. HPLC is fast, simple, specific and reliable over a wide range of sugar concentrations, with no need for derivatization. Sample preparation requires only water-ethanol extraction, followed by a rapid minicolumn cleanup prior to injection into the HPLC system. The method is applicable to baby-foods, cereals, chocolate products, cookies, health-food products, preserves, processed fruits, etc. These products may contain sugars such as fructose, glucose, sucrose, maltose, lactose and others.

Equipment and Materials

Column: stainless steel (30 cm long, 4 mm i.d.) packed with μ Bondapak/carbohydrate.

Mobile phase: acetonitrile-water $(3:1)$; flow rate—2 ml/minute.

Detection: Differential refractometer.

Cleanup microlumn: glass (27.5 cm long, 6 mm o.d.) with $\frac{1}{2}$-inch taper with silanized glass-wool plug and packed with $\frac{1}{8}$-inch Corasil AX and $\frac{3}{4}$-inch Porasil B/C_{18}.

Extraction solution: ethanol-water $(1:1)$.

Standard solutions: mixed solution 1 (1%)—one gram of each sugar is introduced into a 100-ml volumetric flask and diluted to volume with water; mixed solution 2 (0.5%)—50 ml of solution 1 are pipetted into a 100-ml volumetric flask and diluted to volume with water.

Preparation of Sample

Ten grams of the sample are weighed into a volumetric flask (100 ml for $\leq 20\%$ total carbohydrates) and the extraction solution is added to three-quarters of the final volume. The samples are placed for one hour in a sonic bath at ca. 50°C. They are then removed, cooled to room temperature, diluted with water and well shaken to ensure thorough mixing. Samples containing $\geq 8\%$ fat are extracted with chloroform to remove the fat. Those containing $\geq 50\%$ protein are adjusted to pH 4.5 to precipitate the protein.

These special treatments are necessary to prevent overloading of the minicolumn. The samples are centrifuged for 10 minutes at 2000 rpm.

Minicolumn Cleanup

One milliliter of acetonitrile-water solution $(3:1)$ is passed through a prepared minicolumn and discarded. Subsequently, 1 ml of the sample solution is drained through the minicolumn and discarded. A further 1-ml aliquot of the sample solution is added to the column, and the clear eluate is collected in a screw-capped collection tube. The samples are now ready for HPLC.

Procedure

Twenty-five microliters of mixed standard solution 1 or 2 (both before and after minicolumn percolation) are injected into the chromatograph. Recovery of the standard through the minicolumn is determined using peak height measurements.

The individual sugar content in sample is calculated as follows:

$$\text{Sugar, wt}\% = (\text{PH}/\text{PH}') \times (\text{W}'/\text{V}_1) \times (\text{V}/\text{W}) \times 100$$

where PH and PH′ are the peak heights of sugar in the sample and standard, respectively; V is the volume (in milliliters) used to extract the sample (either 100 or 200 ml); V_1 is the final volume (in milliliters) of standard solution; and W and W′ are the weights (in grams) of the sample and standard, respectively.

Standard solutions are injected before the sample and after every 6 sample injections to monitor any change in retention volumes, peak heights and shapes.

If unexpectedly low concentrations of sugars are found in the sample, injection volumes are increased to 50 μl.

The order of elution is as follows: fructose, glucose, sucrose, maltose, lactose, melibiose, raffinose, stachyose.

TLC OF SYNTHETIC SWEETENERS [2]

Artificial or synthetic sweeteners are used as sugar substitutes for dietetic reasons. Until recently, the two most important nonnutritive synthetic sweeteners were calcium or sodium cyclamate (I) and saccharin (II). Dulcin (III), discovered more than a hundred years ago, and P-4000 (IV), discovered in 1940, were found to be very sweet but are, unfortunately, toxic to man.

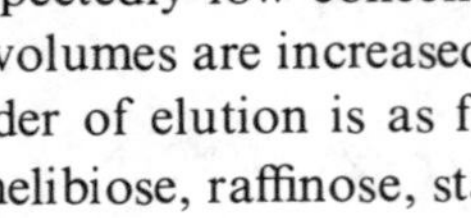

I

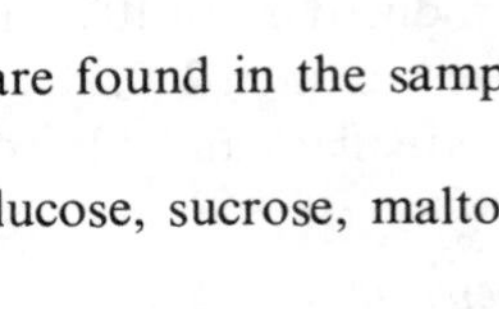

II

III

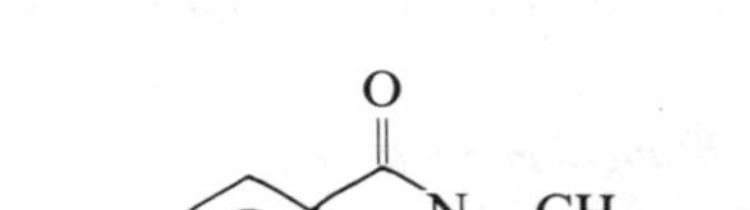

IV

Separation of Calcium Cyclamate (I), Sodium Saccharin (II), Dulcin (III) and P-4000 (IV)

The separation is performed on silica gel G plates. The plate is spotted with dulcin and P-4000, both dissolved in 95% ethanol, and the sodium salt of saccharin and calcium cyclamate, both dissolved in a mixture of equal volumes of 95% ethanol and water. The developing solvent is a mixture of butanol-ethanol-ammonia(28%)-water (40:4:1:9). The developed plates are examined under shortwave (254 mμ) UV radiation, saccharin appearing as a blue fluorescent spot and P-4000 as a dark spot. A solution of 1% chloranil in benzene is sprayed on the plate which is then heated in an oven at 100°C for about 15 minutes. Saccharin and cyclamate appear as white spots on a lavender background. A light spraying with 1% *p*-dimethylaminobenzaldehyde in 10% (w/v) HCl reveals dulcin as a bright yellow spot on a white background, while the P-4000 spot remains lavender.

	R_f
Calcium cyclamate	0.30
Sodium saccharin	0.41
Dulcin	0.75
P-4000	0.84

GLC TEST FOR HONEY ADULTERATION BY HIGH-FRUCTOSE CORN SYRUP [3]

Bound-enzyme technology has resulted in the production of a new industrial sweetener, high-fructose corn syrup (HFCS). It is an inexpensive, highly refined syrup typically containing about 50% glucose, 42% fructose and small amounts of higher saccharides. Honey, on account of its greater complexity (contains at least 10 disaccharides) and limited supply, is a likely target for adulteration by HFCS and possibly also by other cheap, carbohydrate-containing syrups.

The following is a rapid screening method for the detection of HFCS in honey. It is based on the difference between the ratios of the disaccharides maltose and isomaltose in honey and HFCS.

Apparatus

A gas-liquid chromatograph equipped with flame-ionization detector and a 14-foot, $\frac{1}{8}$-inch i.d. stainless steel column packed with 3% OV-17 on 100- to 200-mesh Gas-Chrom Q is used.

Operating conditions: injection port—300°C; detector—300°C; column —240°C; programming—2°C/minute; helium carrier flow rate—50 ml/minute.

Procedure

About 60 mg of honey or syrup are accurately weighed into a 5-ml screw-top vial, followed by 1 ml of pyridine solution containing 1 mg cholestane (internal standard). After the sample is dissolved, 0.9 ml of hexamethyl-disilazane (HMDS) is added and mixed, and then 0.1 ml of trifluoroacetic acid (TFA) is carefully added dropwise. The sample is shaken for 30 seconds and then allowed to stand for 15 minutes with occasional shaking. A homogeneous, clear solution is formed; 5 μl are withdrawn with a syringe and injected into the GLC column.

Determination of Maltose and Isomaltose

Silylated carbohydrates (sucrose, maltose and isomaltose—α and β anomers) are eluted in the following order: sucrose, α-maltose, β-maltose, α-iso-maltose, β-isomaltose, cholestane.

The chromatograms of the standards are then compared with those of honey and HFCS samples.

HPLC DETERMINATION OF THEOBROMINE AND CAFFEINE IN COCOA AND CHOCOLATE PRODUCTS [4]

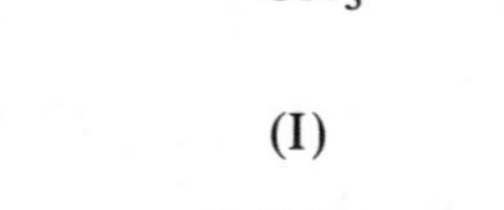

(I)

Caffeine

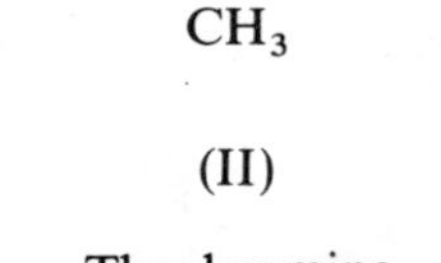

(II)

Theobromine

HPLC is a rapid method for quantitation of caffeine (I), not only in cocoa and materials low in nutritive sweeteners, but also in all types of confectionery and food items containing cocoa solids.

Equipment and Materials

Column: stainless steel (30 cm long, 4.6 mm i.d.) packed with μ Bondapak C_{18}.
Mobile phase: methanol-acetic acid-water (20 + 1 + 79); flow rate— 1 ml/minute.
Detection: UV (280 nm).
Alkaloid standards: alkaloids, caffeine (I) and theobromine (II) are purified by sublimation at atmospheric pressure. Standard solutions are prepared by dissolving 50 μg of alkaloid per milliliter of distilled water.

Procedure

Samples of cocoa (0.6 g), milk chocolate (4.5 g), or chocolate liqueur (1.0 g) are weighed into test tubes equipped with Teflon-lined screw caps. The fat is extracted by shaking twice with 30-ml portions of petroleum ether, centrifuging at 2000 rpm for 10 minutes, and carefully decanting the solvent. The residue is freed of solvent by placing the test tubes in a warm water bath in a hood. The dry test tubes and contents are weighed to determine the weight of the residue. The residue is transferred with distilled water to a weighed Erlenmeyer flask and made up to a volume of 95 ml with water. The Erlenmeyer flask is heated in a hood for 25 minutes at 100°C. The flask is then cooled to 20°C and 100 ml of distilled water are added. The flask is stoppered and shaken thoroughly, and the contents are centrifuged at 2000 rpm for 5 minutes and then filtered through a 0.45-nm membrane filter. The standard solutions are injected into the HPLC column, followed by the aqueous extracts of the samples.

HPLC ANALYSIS OF THE PUNGENT PRINCIPLES OF PEPPER [5]

The alkaloid fraction of black pepper contains a mixture of geometric isomers of isochavicine (1) (*trans-cis*), isopiperine (2) (*cis-trans*), chavicine (3) (*cis-cis*) and piperine (4) (*trans-trans*).

In order to determine the concentration of piperine, the isomer responsible for the pungency of pepper, and to develop a rapid and efficient method for the separation of the isomers of pepper, a sample of pure piperine is photoisomerized and the products are analyzed by HPLC.

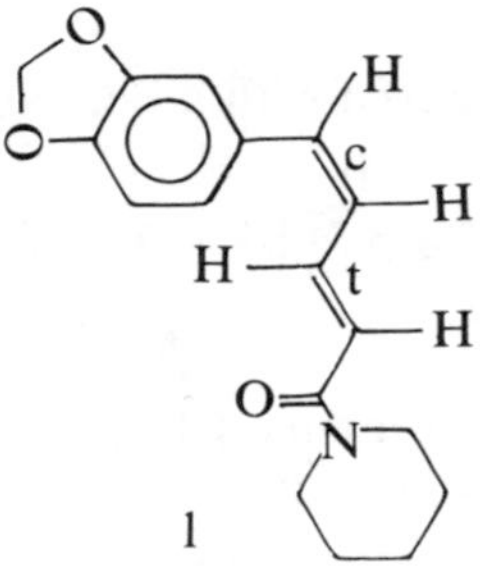

1

Isochavicine

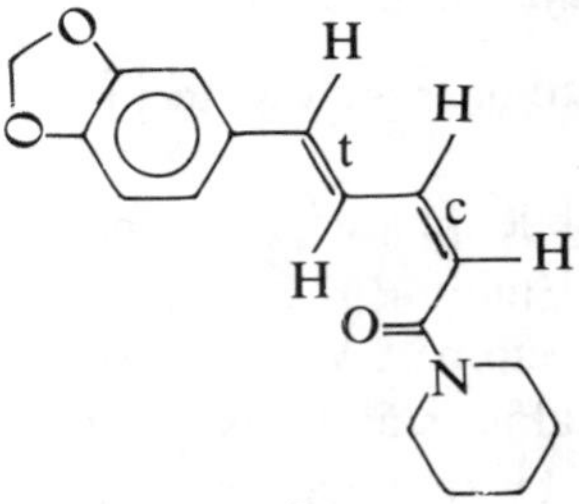

2

Isopiperine

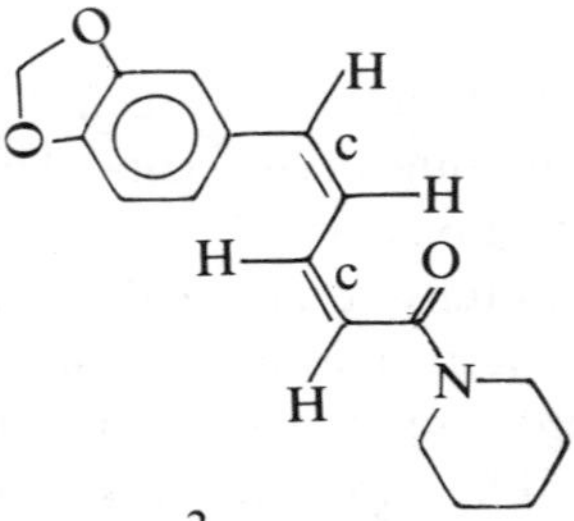

3

Chavicine

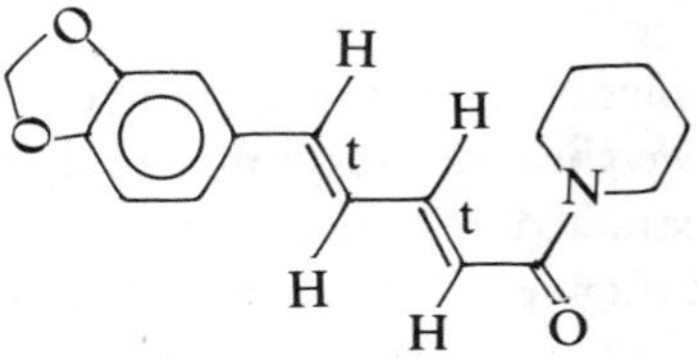

4

Piperine

Complete separation of the isomers is obtained on a nitro-silica gel stationary phase, the compounds being eluted in the following order: chavicine, isochavicine, isopiperine, piperine.

Procedure

Pepper extraction

Pepper is ground in a coffee mill or mortar and extracted for 3 hours with methylene chloride in a soxhlet extractor. Repeated recrystallization of the residue (after removal of methylene chloride) from methanol yields pure piperine, m.p. 127°C.

Photo-isomerization of piperine

Piperine (1 g) dissolved in methanol (100 ml) is irradiated at 350 nm in a photochemical reactor. The photostationary state is reached after about 4 hours. Photo-isomerization also occurs in daylight and takes several days to reach completion.

Nitro-silica gel

Phenyl-silica gel, obtained by reaction of silica gel with phenyltrichlorosilane (10 g), is mixed with sufficient fuming sulphuric acid to obtain a slurry (15 ml). The slurry is heated to 40°C and nitric acid of specific gravity 1.5 (5 ml) is slowly added, keeping the temperature below 50°C. After 5 hours the slurry is poured into 1 liter of cold water. The nitro-silica gel stationary phase is washed until acid-free and then dried at 110°C in an oven.

HPLC

A stainless steel tube (25 cm long, 4.6 mm diameter) is filled with a suspension of nitro-silica gel in carbon tetrachloride. The column is rinsed with methanol and methylene chloride and equilibrated with the eluting solvent (methylene chloride-methanol, 100:4.5). The analysis takes about 15 minutes. The solvent flow rate is 2 ml/minute and the detection wavelength is 280 nm.

p-Bromoacetanilide (1 mg/ml methylene chloride) is used as an internal standard.

Peak No.	Compound	M.P (°C)	λ_{max} (nm)	ε
1	Isochavicine	103	336	12500
2	Isopiperine	86	335	13200
3	Chavicine	(75)	321	10900
4	Piperine	129	343	34100

DETERMINATION OF TOCOPHEROLS AND TOCOTRIENOLS IN FOODS AND TISSUES BY HPLC [6]

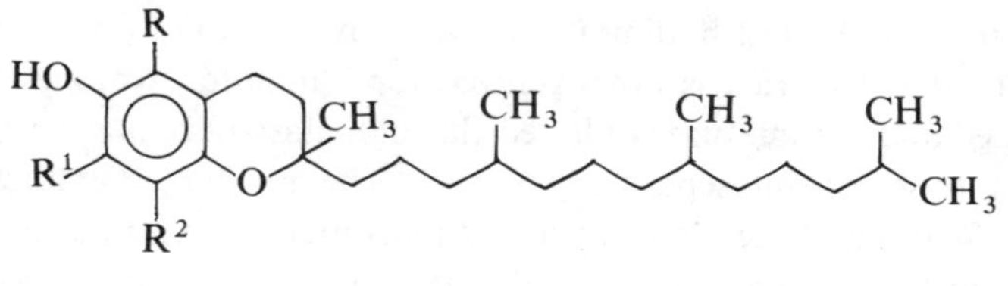

	R	R^1	R^2
α-Tocopherol (5,7,8-Trimethyltocol)	CH$_3$	CH$_3$	CH$_3$
β-Tocopherol (5,8-Dimethyltocol)	CH$_3$	H	CH$_3$
γ-Tocopherol (7,8-Dimethyltocol)	H	CH$_3$	CH$_3$
δ-Tocopherol (8-Methyltocol)	H	H	CH$_3$

	R	R¹	R²
α-Tocotrienol	CH_3	CH_3	CH_3

The earliest methods for determining naturally occurring forms of vitamin E, which consists of eight tocols [four tocopherols (α-, β-, γ- and δ-) and
four corresponding tocotrienols], involved simple colorimetric tests. More
sophisticated procedures have since been developed, involving GLC, TLC,
paper and column chromatography, which can be used to separate and
measure all forms of the vitamin. The following HPLC method is selective
and sensitive to all types of samples.

Apparatus

A 5-μ or 10-μ silica-gel column (25 × 3.2 cm) is used.

The effluent is monitored by means of a spectrophotofluorometer,
equipped with a flow cell and a LC 55 absorption detector (set at 295 nm).

The excitation is set at 290 nm and the emission at 330 nm; a high-pass
filter (310 nm) is placed in the emission beam.

Extraction of Lipids

Samples of foods, such as infant formula, wheat flour, barley and plant
oils, are mixed until uniform. An aliquot (10 g) is weighed and placed for
10 minutes in 100 ml of boiling isopropanol in the cup of a virtis homogenizer. The digest extract is homogenized for 1 minute. Acetone (50 ml) is
then added and the mixture is filtered through glass-fibre paper (Whatman
GF/A) into a 500-ml separatory funnel. The residue is extracted with
acetone (50 ml) and the filter paper and its contents are homogenized with
another portion of acetone (100 ml). After work-up as described above,
hexane (100 ml) is added to the combined acetone extracts and the separatory
funnel is swirled to mix the solvents.

Water (100 ml) is added, the funnel is swirled gently, and the phases are
allowed to separate. The hexane epiphase is transferred to a second separatory funnel and the aqueous hypophase is then extracted twice with

100-ml portions of hexane. The combined hexane extracts are washed twice with water (100 ml) and evaporated under reduced pressure.

Analytical HPLC

The solvent system is 5% diethyl ether in moist hexane; the flow rate is 2 ml/minute. A mixture of standards (α-, β-, γ-, and δ-tocopherols) is injected after every two samples. The volume injected is usually 20 μl.

A solution of barley oil (0.5%), which is a rich source of α-, β- and γ-tocotrienols, is used in conjunction with the tocopherol standards to develop a system capable of separating the four tocopherols and three tocotrienols. Good separations are achieved on 10-μ or 5-μ silica gel in moist hexane containing either 5% diethyl ether or 0.2% isopropanol.

Preparative HPLC

Pure tocotrienols are obtained by preparative HPLC of extracts of rubber latex (a source of α- and γ-tocotrienols) and wheat flour (a source of β-tocotrienol). Lipids from three 10-ml aliquots of rubber latex are extracted as described above. The extracts are evaporated and saponified, by refluxing with 50% aqueous KOH (5 ml) for 5 minutes. The unsaponifiable lipid is then chromatographed in two portions on a 25 × 1 cm diameter column packed with 10-μ or 5-μ silica gel. The column is eluted with 5% diethyl ether in moist hexane (10 ml/minute) and the eluate is monitored at 295 nm. Two fractions are collected, one containing α-tocotrienol and the other γ-tocotrienol. A specimen of β-tocotrienol is obtained by applying the same procedure to whole wheat flour.

DETERMINATION OF FOOD PRESERVATIVES

$$HO-\!\!\left\langle\!\bigcirc\!\right\rangle\!\!-COOR$$

$$R = CH_3, C_2H_5, C_3H_7, C_4H_9.$$

HPLC of food preservatives [7]

Methyl-, propyl-, and butyl-p-hydroxybenzoate esters (parabens) are frequently employed as preservatives in pharmaceuticals, cosmetics and foods. Their widespread use can be attributed to their broad antimicrobial spectrum under acid, alkaline or neutral conditions, combined with their properties of stability, nonvolatility and relative nontoxicity. Various methods have been proposed for the extraction and determination of parabens and saccharin in foods and pharmaceuticals. Most of them are tedious and lead to the formation of emulsions during the process of extraction.

The following method of extraction involves the use of disposable Extrelut (Merck) cleanup columns, which considerably simplifies the isolation of organic compounds from foods, cosmetics and pharmaceuticals. The quantitative determination of the extracted compounds can be performed by GLC, TLC or HPLC.

Equipment and Materials

Column: stainless steel (25 cm long, 4.6 mm i.d.) packed with μ Bondapak C_{18}.
Detection: UV (235 nm).
Cleanup column (see details on p. 77): Extrelut pre-packed columns and refill packs are obtainable from Merck.
Phosphate buffer solution: prepared by dissolving 2.5 g $K_2HPO_4 \cdot 3H_2O$ and 2.5 g KH_2PO_4 in redistilled water and passing the solution through a 0.45 μm filter.
Standard solutions: 25 mg of each of the following compounds are dissolved together in 25 ml methanol and diluted to 50 ml in a volumetric flask: saccharin, benzoic acid, sorbic acid, methyl-*p*-hydroxybenzoate, ethyl-*p*-hydroxybenzoate and propyl-*p*-hydroxybenzoate.

The stock solution is diluted 1:10 for PHB esters alone, the two acids plus saccharin resulting in a concentration of 0.06 μg/μl per compound.

Procedure

Direct analysis
Juices (such as orange juice), wines and other aqueous media generally require no cleanup. Filtration is, however, recommended for eliminating any particulate matter.

Samples containing fats
Samples with a high fat content, such as margarine, should be extracted according to the following procedure: a 5-g sample is dissolved in 50 ml of

diethyl ether and extracted twice with 10 ml of 0.1 N sodium hydroxide solution in a separatory funnel. The basic aqueous extracts are acidified with 1 ml of 5 N sulphuric acid in a 50-ml volumetric flask and diluted to volume with methanol.

Extrelut cleanup

This method is suitable for most other foodstuffs, such as cheese, cakes, yoghurts and other samples, that tend to form emulsions during extraction. The pre-packed or refilled Extrelut column in a plastic tube consists of wide-pore kieselgur of grainy structure and is characterized by a high pore volume. The tube is capped at both ends with screens containing a filter disc. A sample (5 to 20 g) is homogenized for 3 minutes in 50 ml of 0.5 N sulphuric acid using a beaker and a high-speed blender. The homogenate is transferred quantitatively to a 100-ml volumetric flask and diluted to volume with water. Twenty milliliters of this pre-treated sample are pipetted onto the Extrelut column and allowed to infiltrate for at least 15 minutes. It is important to replace the original filter disc by some glass wool when working with samples of thicker consistency.

The absorbed preservatives and saccharin can now be eluted with 350 ml of chloroform-isopropanol (9:1). The eluate is collected in a 500-ml round-bottomed flask and carefully evaporated almost to dryness under vacuum. The residue is transferred with methanol to a 10-ml volumetric flask and diluted to volume with methanol. The filtered extract is now ready for analysis.

Determination of the five preservatives

In order to analyze the p-hydroxybenzoate (PHB) esters together with the two acids, the following procedure is adopted. The column is initially eluted with 20% methanol in phosphate buffer, followed by a linear gradient from 20% to 80% methanol at a flow rate of 2 ml/minute.

PHB esters alone

Elution is carried out with 60% methanol in phosphate buffer; flow rate — 1.2 ml/minute.

The following food samples may be used: milk chocolate, marmalade, ketchup, fruit cocktail, canned vegetables, mixed pickles, cake, cheese.

GC of food preservatives

Equipment and Materials

TMS derivatives of benzoic and sorbic acid and the methyl ethyl and propyl esters of p-hydroxy benzoic acid are prepared, and 1 ml of internal standard (derivative of methyl gallate) is added.

A 3-m stainless steel column packed with 3% SE-30 on 100- to 120-mesh Aeropak is used. Temperature programming: from 90°C to 290°C at a rate of 8°C/minute.

Preparation of Samples

For the preparation of samples of marmalades, mustard, mayonnaise, sardines in oil and margarine, 10 g of the sample food and 60 g of sand are thoroughly mixed in an Erlenmeyer flask. After the addition of 3 ml sulfuric acid (25%), the mixture is extracted three times with 30 ml of ether by agitating the ground-glass-stoppered flask vigorously for at least 1 minute. The preservatives are then extracted from the ether by two 20-ml portions of 0.1 N NaOH. After each alkaline extraction, 10 ml of saturated NaCl solution are added. The aqueous alkaline layer is neutralized with HCl (1:3). The preservatives are then extracted from the aqueous phase with one 100-ml and four 50-ml portions of chloroform. The extract is dried over anhydrous sodium sulfate. The chloroform extract is evaporated to 1 ml. One milliliter of internal standard (14 mg/ml methyl gallate in pyridine) is added. The 2 ml are then concentrated to approximately 1 ml and 0.2 ml N,O-bis (trimethylsilyl) acetamide (BSA) is added. This mixture is then heated under reflux for 15 minute at 60°C and the resultant trimethylsilyl (TMS) derivatives are then injected into the gas chromatograph.

The order of emergence of peaks from the column is as follows: sorbic acid, benzoic acid, methyl-p-hydroxybenzoic acid, ethyl-p-hydroxybenzoic acid, propyl-p-hydroxybenzoic acid, as their TMS derivatives.

TLC of food preservatives [9]

Procedure

Preparation of plates

A mixture of 15 g silica gel GF_{254}, 7.5 g cellulose $MN300F_{254}$ and 70 ml of water is applied on five 20 × 20 cm plates with an applicator. The plates are air-dried and then activated by heating at 110°C for 30 minutes.

Solvent and development

A mixture of light petroleum (b.p. 40–60°C)-carbon tetrachloride-chloroform-formic acid-acetic acid (50:40:20:8:2) is shaken in a separatory funnel, the two layers are allowed to separate, and the upper layer only is used as the mobile phase. The plate is eluted twice to a length of 15 cm with the same solvent. A mixture of nine preservatives, namely, benzoic acid, sorbic acid, salicyclic acid, dehydroacetic acid, bromoacetic acid, propyl-p-hydroxybenzoate, ethyl-p-hydroxybenzoate, methyl-p-hydroxybenzoate and p-hydroxybenzoic acid, can be separated.

Detection

For the detection of benzoic acid, a solution of 4.5 ml H_2O_2 (30%), 4.5 ml water and 1 ml saturated $MnSO_4$ is used, followed by a 0.3% aqueous solution of $FeSO_4$; for the detection of sorbic acid—a solution of 5 ml $K_2Cr_2O_7$ (0.5%) and 5 ml 0.3 N H_2SO_4, followed by a saturated solution of thiobarbituric acid; for the detection of salicylic acid—a 0.1% aqueous solution of $FeCl_3$; and for the detection of dehydroacetic acid—a 3% aqueous solution of $TiCl_3$ or a 0.1% aqueous solution of $FeCl_3$. For the detection of bromoacetic acid, a mixture of three volumes of Phenol Red [24 mg Phenol Red in 2.4 ml NaOH (0.1 N) made up to 100 ml with acetone] and one volume of a CH_3COONa solution (6 g CH_3COONa, 3 ml CH_3COOH and water made up to 100 ml) is sprayed on the chromatogram, followed by a spray of a Chloramine T solution (25 mg Chloramine T in 15 ml of water-acetone, 1:1). For the detection of esters and free p-hydroxy-benzoic acid, a 10% NaOH solution, a 2% alcoholic solution of aminoantipyrine and an 8% aqueous solution of $K_3Fe(CN)_6$ are used. The esters are hydrolyzed by spraying with the NaOH solution. The plate is heated for 5 minutes at 80°C, sprayed again with distilled water, and heated for another 5 minutes. Spraying with the aminoantipyrine and $K_3Fe(CN)_6$ solutions gives red to red-brown spots. Their R_f values are summarized below.

	R_f
Benzoic acid	0.70
Sorbic acid	0.64
Salicylic acid	0.56
Dehydroacetic acid	0.50
Bromoacetic acid	0.30
Propyl-p-hydroxybenzoate	0.25
Ethyl-p-hydroxybenzoate	0.20
Methyl-p-hydroxybenzoate	0.13
p-Hydroxybenzoic acid	0.06

TLC OF WATER-SOLUBLE VITAMINS [10]

Water-soluble vitamins are heterogeneous compounds, some of which are chemically unstable. TLC is a suitable method for analysis of these vitamins, which may be separated on rice-starch-coated plates using ninhydrin for detection.

Procedure

The thin layers of rice-starch-coated plates (20 × 20 cm) are prepared by suspending 18 g of rice starch and 2 g of gypsum in 20 ml of 96% ethanol (to prevent the formation of bubbles which appear when rice starch is suspended in water) and then in 40 ml of distilled water. The resulting suspension is applied with a TLC-applicator to clean, fine-glass plates (0.2 mm thickness). The plates are finally dried at room temperature.

Solutions of eight water-soluble vitamins are prepared by dissolving 2–15 mg of each per milliliter of distilled water (see Table 5). Spots of 1 μl of the vitamin solution are applied, using a micropipette, on the thin-layer rice-starch plate. The chromatoplate is developed by the ascending technique in a glass chamber containing 50 ml of n-butanol-acetic acid-water-pyridine (40:10:50:2) solvent mixture. The chromatogram is run in the dark at room temperature without prior saturation of the chamber with the solvent. Developing time for the chromatogram is about five hours for a solvent front of 15 cm.

Table 5
R_f Values and Color of Spots of Water-Soluble Vitamins

Vitamin	R_f	Ninhydrin Reagent	UV (254 nm)	UV (365 nm)
Thiamin-HCl (B$_1$)	0.42	Light yellow	—	Dark
Riboflavin (B$_2$)	0.18	Dark yellow	Yellow	Yellow
Pyridoxin-HCl (B$_6$)	0.58	Reddish	Blue	Blue
Cyanocobalamin (B$_{12}$)	0.79		—	Dark
Ascorbic acid (C)	0.27	Red violet	Dark	Dark
Nicotinamide	0.74	Light blue	Dark	Dark
Pantothenic acid (Ca-salt)	0.69	Violet	—	—
p-Aminobenzoic acid (PABA)	0.82	Orange	Yellow	Yellow

Detection

The ninhydrin reagent is prepared by dissolving 0.5 g ninhydrin in 100 ml methanol. The developed and dried chromatogram is heated in an oven for 30 minutes at 160°C. After cooling, the chromatogram is sprayed with ninhydrin solution and the chromatogram is again heated for about 10 minutes at 80°C. Colored spots of the vitamins appear under UV light at 254 nm and 365 nm.

HPLC OF ANTIOXIDANTS [11]

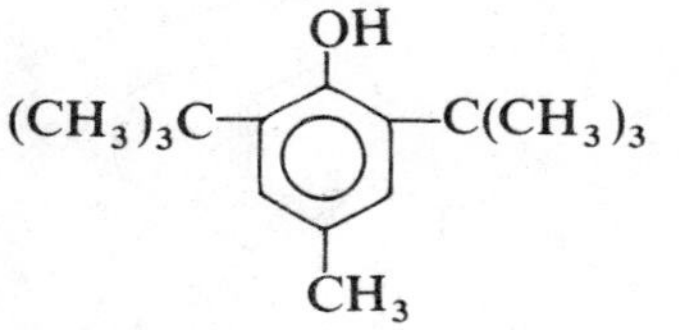

Butylated hydroxytoluene 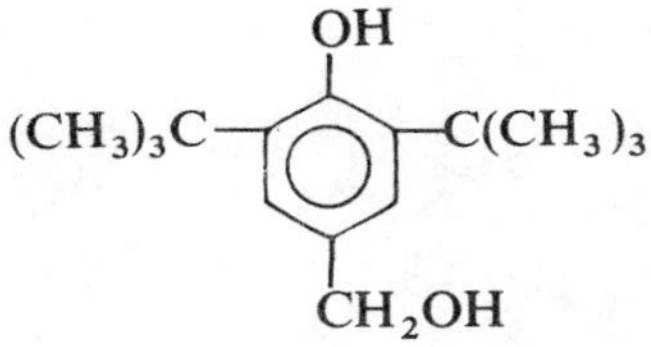   4-Hydroxymethyl-2,6-di-*t*-butylphenol

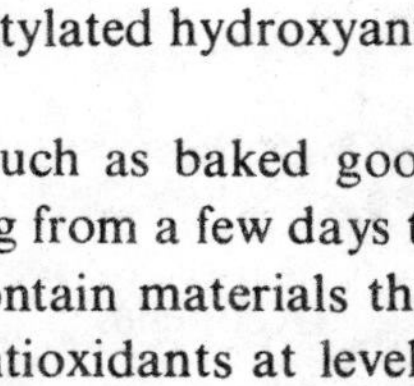

Butylated hydroxyanisole

Many consumer products, such as baked goods, drug formulations, etc., are stored for periods ranging from a few days to several years prior to use. Since these products may contain materials that degrade via oxidation, it is often necessary to add antioxidants at levels carefully chosen so as to maximize product stability while minimizing side effects such as toxicity. In the following experiments three antioxidants are separated by HPLC.

Equipment and Materials

Column: an amino column (25 cm in length, 4.6 mm i.d.).
Mobile phase: 90% hexane/10% isopropanol; flow rate—1.5 ml/minute.
Detection: UV (254 nm).

The elution order is as follows: butylated hydroxytoluene (BHT); 4-hydroxymethyl-2,6-di-*t*-butylphenol, butylated hydroxyanisole (BHA).

REVERSED-PHASE HPLC OF CAROTENES IN TOMATOES [12]

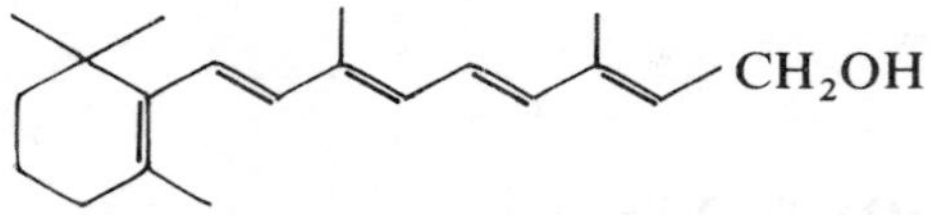

Retinol

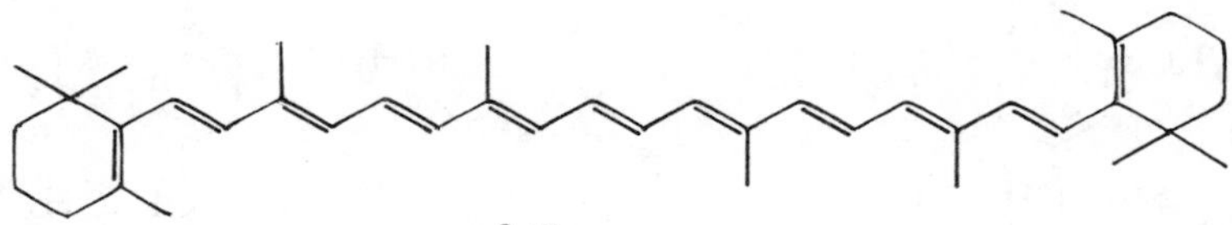

β-Carotene

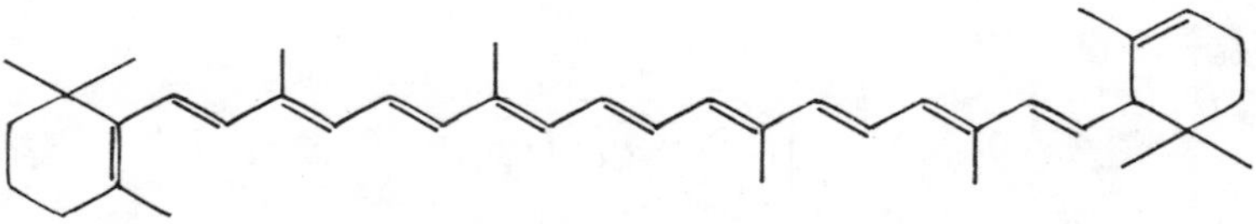

α-Carotene

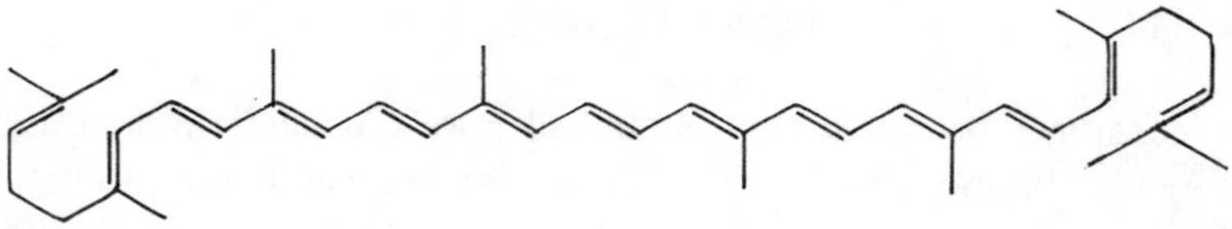

Lycopene

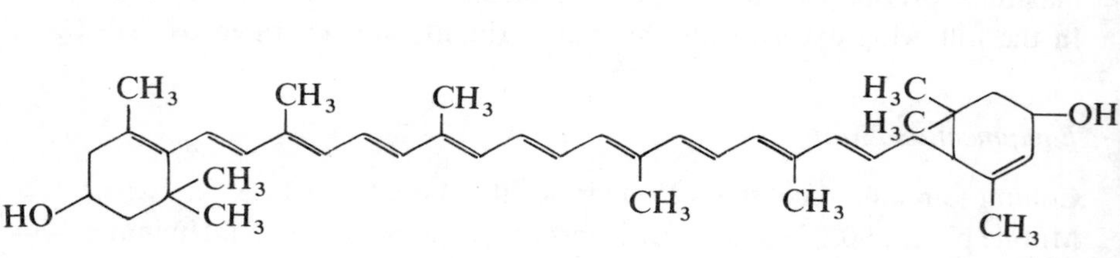

Xanthophyll

Differences in the biopotency of carotenoids as vitamin-A precursors result from their individual structures. The β-ring present in retinol is essential for their activity. β-Carotene, having two such rings, is considered 100% vitamin-A-active. α-Carotene is only one half as potent, while acyclic carotenes also occurring naturally in foods, such as lycopenes, are inactive.

The methods used for the isolation of carotenoids, such as open-column chromatography, are not only time consuming and call for long-term exposure of the carotenoids to oxygen and light, but also often fail to resolve the most potent vitamin-A precursor, all-*trans*-β-carotene, from its less active geometrical isomer, α-carotene; these methods and TLC lack in reproducibility and accurate quantitation. GC cannot be used with thermally labile carotenoids.

Carotenoids are quantitatively separated on reversed-phase packings, which are neutral and unaffected by the presence of water or changes in the mobile phase. The following procedure describes the separation of carotenoids in tomato samples by reversed-phase HPLC.

Procedure

Extraction

Tomato samples (each weighing ca. 150 g) are cut into small pieces and homogenized under a stream of nitrogen in acetone for 1 to 2 minutes in a blender. The homogenate is filtered through a sintered-glass funnel under reduced pressure and the residue is recovered for extraction. The procedure is repeated until complete extraction of all pigments is achieved. The acetone extract is then added to an equal volume of freshly distilled petroleum ether in a separatory funnel, mixed and diluted with water. Upon the formation of two layers, the lower, aqueous phase is re-extracted once with petroleum ether and the combined petroleum ether solutions are washed three times with water to remove acetone.

Saponification and removal of sterols

Saponification is generally necessary in carotenoid analysis to remove unwanted lipid material which could interfere with the chromatography of compounds of interest. Extracts are evaporated to dryness using a rotary evaporator, and a solution of 15% KOH in methanol is added to the round-bottomed flask. The alkaline mixtures are left in the dark for 14 hours at room temperature. They are then gradually added to freshly distilled petroleum ether in a separatory funnel. Water is slowly poured into the funnel so as not to form an emulsion. When two phases appear, the lower, aqueous phase is drawn off and extracted three times with fresh volumes of petroleum ether. The ethereal solutions are then bulked in a separatory funnel and washed free from alkali by repeated additions of water, the resultant aqueous layers being discarded. Each saponified extract is then concentrated to 100 ml in a rotary evaporator and stored under nitrogen in a volumetric flask in a freezer and protected from light.

The different samples are kept in the freezer overnight at $-10°C$, during which period the sterols are precipitated on the bottom of the flasks.

Aliquots for HPLC

Both 50-μl and 10-μl samples in petroleum ether are injected into the HPLC column, the former mainly for determination of the amount of β-carotene and the latter for the determination of lycopene. Concentrations of each compound are determined from the slope of the calibration plots in which the peak is plotted against the amount injected.

Standard solutions of α- and β-carotene are prepared by dissolving 0.2–0.5 mmoles of carotene in 1 liter of petroleum ether; lycopene solution is prepared by dissolving 0.3 mmoles of lycopene in 1 liter of dichloromethane. The separation may be carried out on a stainless steel column (25 cm long, 4.6 mm i.d.) filled with Partisil-10/ODS-2 or Partisil-5/ODS, chloroform in acetonitrile 8.5% (2 ml/minute) serving as the mobile phase.

The early eluted peak of the tomato pigments may be a xanthophyll, followed by lycopene and β-carotene; a peak which elutes between lycopene and β-carotene may be due to γ-carotene. α-Carotene is not prominent in any sample. The major components of tomato extracts, lycopene and β-carotene, are present in a ratio of ca. 9:1. These compounds are identified by comparing both the retention times and the visible spectra (380–600 nm) of the peaks of the extracts with those of the standard solutions.

TLC OF CAROTENOID PIGMENTS IN ORANGES [13]

The complete analysis of carotenoid pigments involves extraction, saponification, separation by chromatographic methods and, finally, identification of each pigment. Using TLC, the preliminary fractionation into hydrocarbons, monols, diols and polyols is effected on silica-gel plates developed with the solvent system acetone-petroleum ether (3:7). Each group can be further separated into individual carotenoids by rechromatography on MgO-kieselgur (1:1) plates using the above solvent system and increasing the amount of acetone according to group polarity.

Procedure

Extraction

Orange peels (20 g) or other material are homogenized—cooling with ice, in an Ultra Turrax homogenizer—with acetone in the presence of butylated hydroxytoluene (BHT) as antioxidant and calcium carbonate as neutralizing agent. After filtering under reduced pressure, the residue is re-extracted with the same solvent until colorless. The acetonic extract is mixed with an equal volume of diethylether (peroxide-free), and sufficient saturated saline solution is added to form two layers. The upper layer (epiphase) contains all the pigments. It is washed several times with distilled water.

Saponification and removal of sterols

The diethylether pigment extract is evaporated to dryness in vacuum (at 40°C), absolute ethanol being added to remove the water. The residue is extracted with diethylether (2–3 ml), and an equal volume of 10% KOH in ethanol is added. The mixture is kept for 2–3 hours in a nitrogen atmosphere at room temperature. After diluting with ether, the ethereal solution is washed free of alkali and evaporated in vacuum. The amount of total carotenoids is determined by spectroscopy. The nonsaponifiable matter is dissolved in a minimum volume of methanol and kept overnight in a freezer at −20°C, during which period the sterols precipitate. They are removed by centrifugation in a refrigerated centrifuge.

TLC

After evaporation of the methanolic supernatant solution under reduced pressure, the residue is dissolved in a few drops of chloroform and petroleum ether. The extract is applied as a line on a silica-gel plate (0.4 mm thick) which is developed with 30% acetone in petroleum ether. When the upper band of carotenes reaches a distance of 2 cm from the top of the plate, it is rapidly scraped off into a beaker containing acetone. The plate is further developed with 30% acetone in petroleum ether until a good separation of all zones is obtained.

Each band is further rechromatographed on thin layers of MgO-kieselgur (1:1). The less polar band is developed with 2–5% acetone in petroleum ether. Thus, phytofluene and α-, β- and ζ-carotenes are readily separated and identified by their visible color: phytofluene—blue-green fluorescence in UV light; α-carotene—yellow; β-carotene—orange; and ζ-carotene—lemon yellow.

The monol group is developed with 10% acetone in petroleum ether, and the more polar groups of diols and triols—with 30% acetone in petroleum ether.

Identification of the pigments and colorless polyenes is made on the basis of their chromatographic and spectroscopic properties. The absorption spectrum of each pigment is recorded in a DB Spectrophotometer in the range of 220 to 550 nm (see Table 6).

Table 6
Visible Absorption Maxima of Carotenoids
in Ethanol (nm)

Phytofluene	327, 348, 367
α-Carotene	420, 443, 372
β-Carotene	425, 450, 475
ζ-Carotene	378, 400, 424
OH-α-Carotene	418, 440, 468
Cryptoxanthin	425, 450, 475
Lutein	420, 443, 472
Zeaxanthin	425, 450, 478
Anteraxanthin	418, 444, 470
Luteoxanthin	400, 424, 448
Violaxanthin	416, 438, 468
Citraurin	454

For some functional groups, chemical tests, like the epoxide test and the carbonyl-reduction test, may be used.

HPLC AND TLC OF SYNTHETIC ACID-FAST DYES IN ALCOHOLIC PRODUCTS [14]

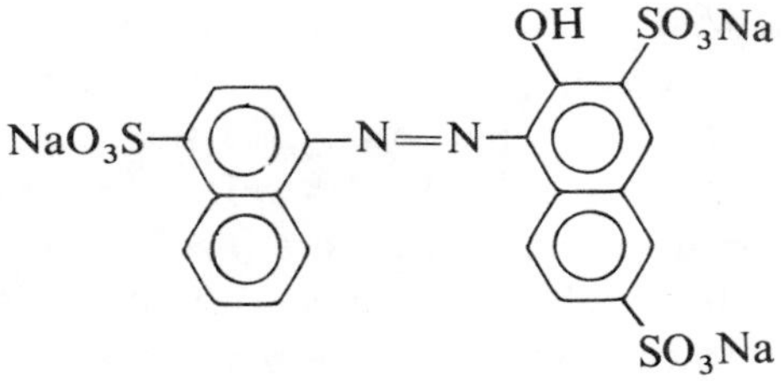

Amaranth

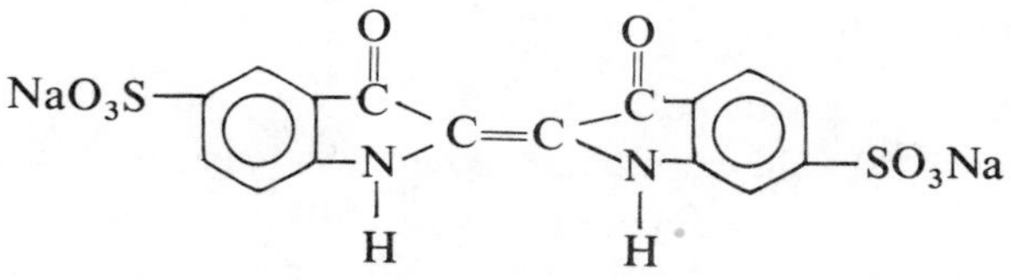

Indigotine

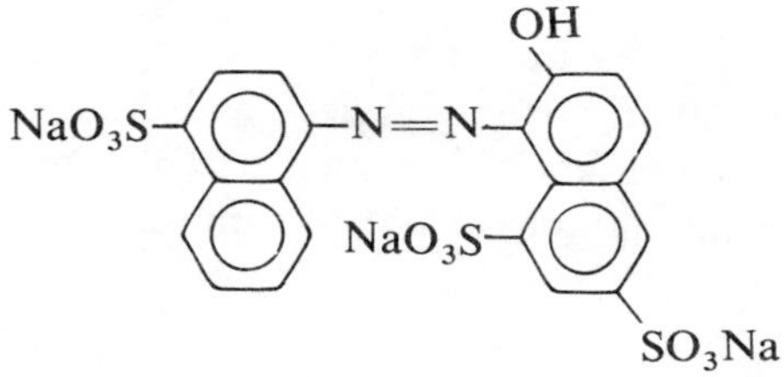

Ponceau 4R

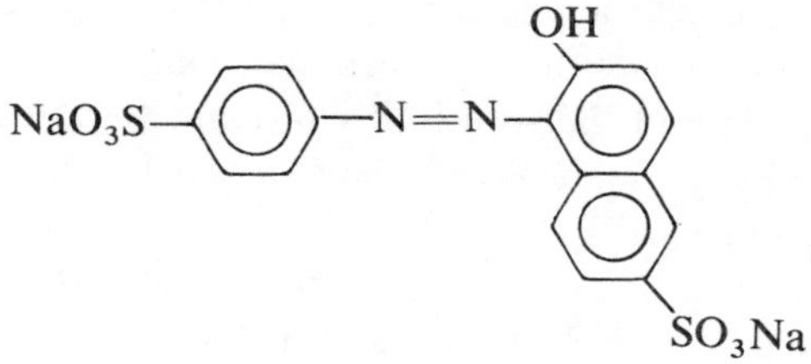

Sunset Yellow FCF

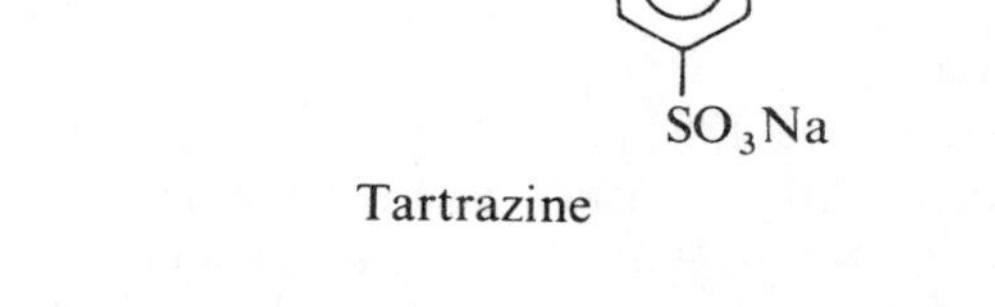

Tartrazine

Synthetic acid-fast dyes, such as found in wines and flavor samples, are extracted with wool yarn and analyzed by HPLC and TLC. The two methods are compared for sensitivity and resolution, and HPLC is found to be quicker and more reliable than TLC.

Equipment and Materials

Glass tanks: $10 \times 12 \times 4$ inches, with lids.

Thin-layer plates: 20×20 cm glass plates, precoated with 0.10 mm cellulose.

Dyes: Amaranth, Erythrosine, Ponceau 4R, Brilliant Blue, Indigotine, Fast Green, Tartrazine, Sunset Yellow FCF. (See above some of the structural formulae.)

Separate solutions of dyes are prepared as follows: each dye (0.1 g) is placed in a separate 100-ml volumetric flask and diluted to volume with water-ethanol (9:1). In addition, mixed standards are prepared by dissolving 0.1 g of each dye and diluting to 100 ml with water-ethanol (9:1).

Preparation of Sample

Flavor extract (25 ml) or alcoholic beverage (50 ml) is placed in a 250-ml beaker. Hydrochloric acid is added until the pH is adjusted to 2. To this solution are added 12 inches of wool yarn and boiling chips. The beaker is placed on a hot plate and the solution is brought to boil. Boiling is continued until the original volume is reduced to one half. The wool is removed from the beaker and rinsed thoroughly with cold water. If the wool is white, no synthetic acid-fast dyes are present. If color is present, the wool is washed in a beaker with 25 ml ammonium hydroxide solution (10%). After 15 minutes, the wool is removed from the beaker, the dye solution is squeezed out, and the wool is discarded. The beaker is then heated on a hot plate until the volume is reduced to 2 ml (avoid charring!).

Solvent system for TLC

The solvent system has the following composition: ethyl acetate-*n*-butanol-pyridine-water (1:1:1.2:1). The plates are developed until the solvent front ascends 10 cm.

HPLC (operating conditions)

Column: stainless steel (25 cm long, 4.6 mm i.d.) packed with RP8 (C_8 chemically bonded to 10-μm particle size silica gel).

Mobile phase: 0.01 M KH_2PO_4 in deionized water and methanol; flow rate—2 ml/minute. The solvent program maintains the mobile phase at 10% methanol for 3 minutes and then the concentration is increased linearly for 15 minutes to a final concentration of 90% methanol.

Detection: UV: 290 nm.

GLC DETERMINATION OF CHOLESTEROL AND OTHER STEROLS IN FOODS [15]

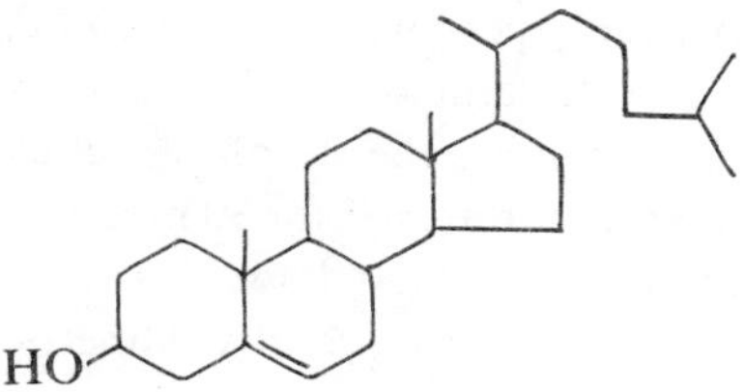

Cholesterol 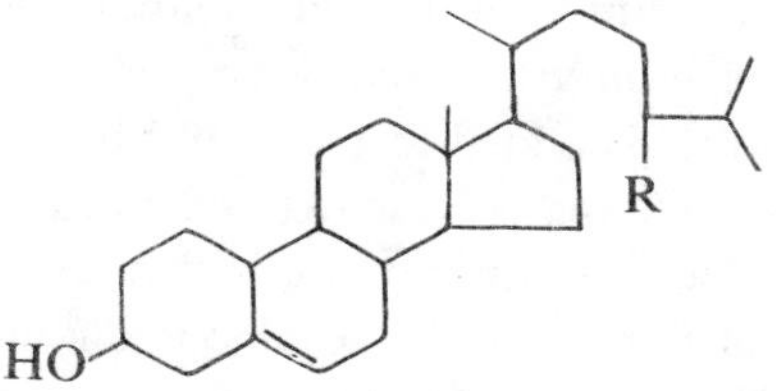   Phytosterol (R = alkyl)

The importance attached to knowledge of the presence and concentration of sterols in foods and to the differentiation between cholesterol present in foods of animal origin and phytosterols present in vegetables and other foods of plant origin has led to the use of GLC for steroid analysis.

Cholesterol is frequently accompanied by α-tocopherol, the latter being inseparable from the former by GLC. However, it is possible to overcome this difficulty by analyzing their derivatives by GLC. In the following method the food samples (such as mayonnaise, doughnuts, custard pie, beef, dry milks, etc.) are saponified and the extracted unsaponifiables are derivatized as butyrates or acetates prior to GLC. This method also makes possible the separation and identification of fatty acids present in the food.

Equipment and Materials

Use is made of a gas-liquid chromatograph equipped with a hydrogen-flame-ionization detector and a glass column (2 m long, 4 mm i.d.) packed with 1% SE-30 on 100- to 120-mesh Gas-Chrom Q.

Operating conditions: temperature of column—250–265°C; injector temperature—300–315°C; detector temperature—300–315°C.

Stock solution

Pure cholesterol (500 mg) is transferred to a 500-ml volumetric flask and diluted to volume with n-hexane. If several sterols are to be determined simultaneously, the appropriate stock solutions are prepared in a manner analogous to that of cholesterol.

Working solutions: Preparation of steryl butyrates or
acetates

The stock solution (10 ml) is transferred to a 50-ml round-bottomed
flask. n-Hexane is evaporated to dryness using a stream of nitrogen. Benzene
(10 ml) and absolute ethanol (2 ml) are added and evaporated to dryness on
a steam bath using a stream of nitrogen. A freshly prepared solution (4 ml)
of n-butyric anhydride-pyridine (2 + 1) (or acetic anhydride, 2 + 1, mixture) is added to the round-bottomed flask containing the sterols. A reflux
condenser is attached to the flask and the mixture is refluxed for 30 minutes.
The reaction mixture is then evaporated almost to dryness under a stream
of nitrogen. The residue is quantitatively transferred to a 5-ml volumetric
flask using n-hexane and diluted to volume with n-hexane.

Hydrolysis of Sample

The homogeneous sample to be analyzed is prepared by grinding, homoge-
nizing, ball-milling or taking a measured aliquot.

It is extracted with chloroform-methanol (2 + 1) (using a volume 20
times that of the sample) and filtered through a filter paper into a separatory
funnel. The organic phase is separated and washed with water. The aqueous
layer is extracted twice more with the chloroform-methanol mixture, and
the organic phase is combined with the first extract. The solvents are evaporated *in vacuo* on a water bath (50°C).

The residue is heated under reflux with 5% methanolic KOH solution
for 2 hours. Methanol is distilled off and replaced by an equal volume of
water. The solution is transferred to a separatory funnel and the nonsaponifiable fraction (NSF) which contains the sterols is extracted several
times with small portions of ether.

Evaporation of the ether leaves a semi-solid residue which is converted
into steryl butyrates or acetates as described above.

The aqueous alkaline layer yields free fatty acids (FFA) upon acidifica-
tion with hydrochloric acid. The acids can be converted to methyl esters
and analyzed by GLC on a DEGS column.

The steryl esters are quantitatively transferred to a tared flask using
petroleum ether, which is evaporated *in vacuo* under a stream of nitrogen.
The residue is weighed, dissolved in petroleum ether, transferred to a
100-ml volumetric flask, and diluted to volume with petroleum ether.

Approximately 2–5 μl of this solution are injected into the gas chromatograph and the chromatogram is compared with that of standard sterol
derivatives.

HPLC AND TLC DETERMINATION OF AFLATOXINS IN CORN [16]

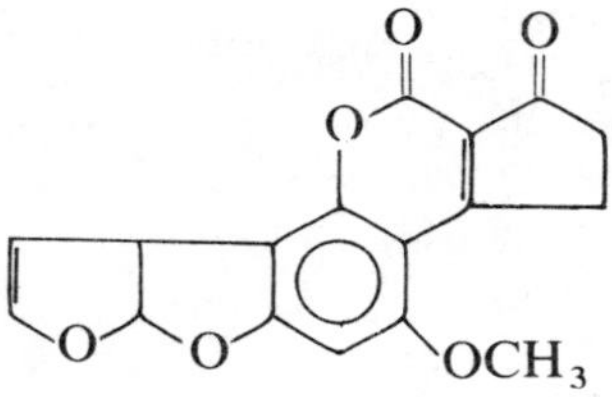

Aflatoxin B$_1$

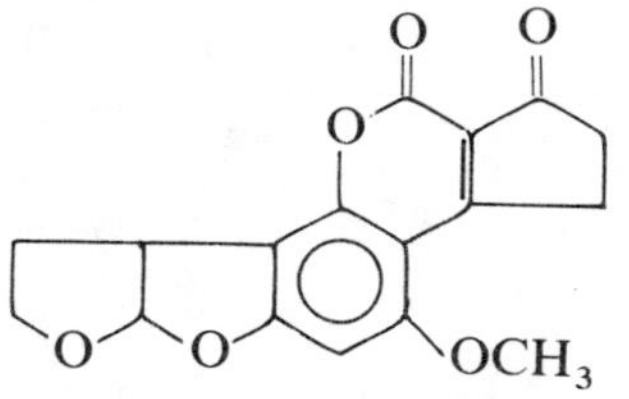

Aflatoxin B$_2$

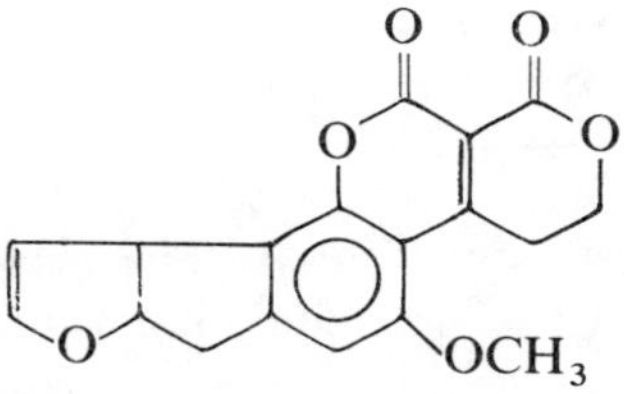

Aflatoxin G$_1$

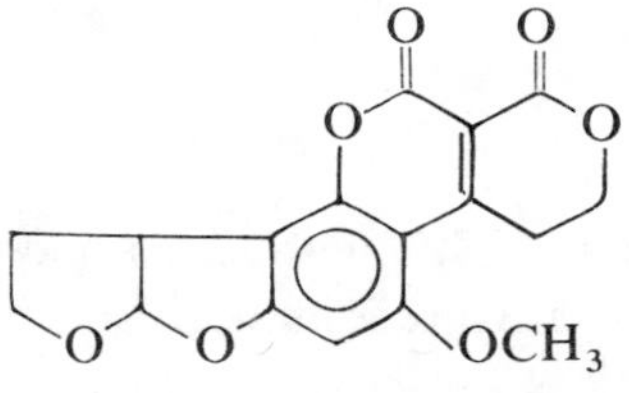

Aflatoxin G$_2$

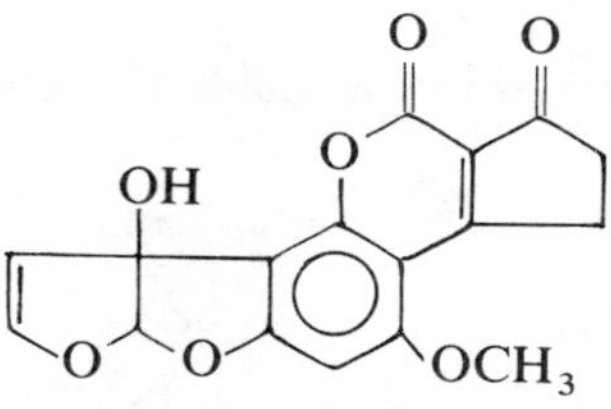

Aflatoxin M$_1$

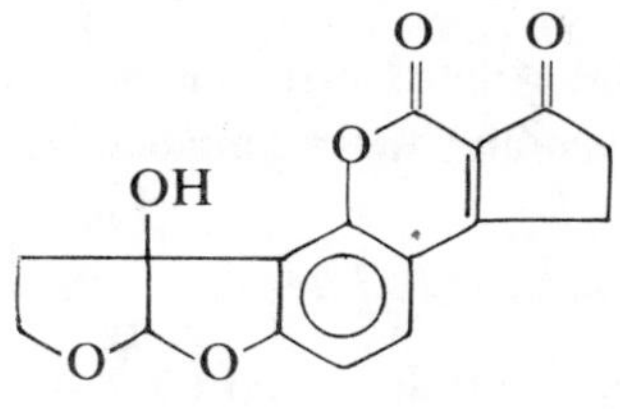

Aflatoxin M$_2$

Aflatoxins are a closely related group of secondary fungal metabolites which have been shown to be mycotoxins. They are produced by *Aspergillus flavus* and were found to be the cause of turkey X disease. Aflatoxins have been reported to occur naturally in peanuts, peanut meal, cottonseed meal, corn, oats, rye, buckwheat, rice and dried chili peppers.

These condensed steroids are characterized as aflatoxins B$_1$, B$_2$, G$_1$, G$_2$, M$_1$, and M$_2$ (M$_1$ and M$_2$ are found as toxins in milk). These six aflaxtoxins are easily identified by their fluorescence under ultraviolet light (see Table 7), R_f values and chemical structures.

Table 7

Aflatoxin	Fluorescence	UV_{max} (ethanol)
B_1	Blue	223, 265, 362 nm
		(25,600, 13,400, 21,800)
B_2	Blue	265, 363 nm
		(11,700, 23,400)
G_1	Green	243, 257, 264, 362 nm
		(11,500, 9,900, 10,000
		16,100)
G_2	Green-blue	265, 363 nm
		(9,700, 21,000)
M_1	Blue-violet	226, 265, 357 nm
		(23,100, 11,600, 19,000)
M_2	Violet	221, 264, 357 nm
		(20,000, 10,900, 21,000)

The following HPLC method utilizes methanol-1% NaCl (4 + 1) to extract aflatoxins from corn, zinc acetate-NaCl to precipitate pigments, and a small (2 g) silica-gel column to clean up the extracts. The aflatoxins are resolved by normal-phase HPLC on microparticulate silica gel using a water-saturated chloroform-cyclohexane-acetonitrile solvent and detected by fluorescence on a silica gel-packed flowcell.

Note: **Aflatoxins are highly toxic and great caution should be exercised when handling these compounds.**

Equipment and Materials

Column: stainless steel (30 cm long, 4 mm i.d.) packed with Porasil silica gel (10 μm).

Mobile phase: the elution solvent is prepared from water-saturated chloroform-cyclohexane-acetonitrile (25 + 7.5 + 1) with added 1.5% absolute ethanol or 2% 2-propanol. The amount of alcohol can be varied to obtain optimum resolution with different silica-gel columns.

Detection: fluorescence (360 nm excitation) or UV (360–365 nm).

Aflatoxins standards: prepared by dissolving 0.5 μg B_1 and G_1 and 0.15 μg B_2 and G_2 per milliliter of elution solvent. The same concentrations in benzene-acetonitrile (98 + 2) are used for TLC.

Zinc salt solution: zinc acetate dihydrate (150 g) and NaCl (150 g) are dissolved in water, glacial acetic acid (1 ml) is added, and the solution is diluted to 1 liter with water.

Silica gel G: silica gel (70- to 230-mesh) is dried for 1 hour at 105°C, 1% water is added, and the mixture is equilibrated overnight in an airtight container.

Extraction

Finely ground corn (50 g) is weighed into a 1-liter blender jar, and 10% sodium chloride (50 ml) and methanol (200 ml) are added. The mixture is blended for 1 minute at a low speed and for 3 minutes at a high speed. It is then filtered under reduced pressure on a Buchner funnel fitted with a circle of rapid filter paper, overlaid with a thin layer of filter aid.

Precipitation and Partition

The filtrate (100 ml) and zinc salt solution (100 ml) are transferred to a 250-ml beaker, stirred and left at room temperature for 5 minutes for coagulation. Filter aid (10 g) is added and the contents of the beaker are stirred and then filtered through folded fluted filter paper; at least 100 ml of the filtrate are collected. The filtrate (100 ml) is transferred to a separatory funnel and shaken vigorously with methylene chloride (25 ml) for 1 minute. After separation of the phases, the lower phase of methylene chloride is drained through a column containing a 4-cm layer of Whatman cellulose powder, topped with 2 cm of anhydrous granular sodium sulfate. The column is washed with 25 ml methylene chloride. The solvent is evaporated cautiously on a steam bath. Overheating of the dry extract can destroy aflatoxins.

Column Cleanup

The column is prepared by slurrying 2 g silica gel-60 in 5 ml ether-hexane (3:1). The slurry is poured into the column. When the gel settles, 1.5 g anhydrous granular sodium sulfate are added, and the solvent is drained to the top of the sodium sulfate. The dry sample extract in 2 ml methylene chloride is poured into the column and drained to the top of the sodium sulfate. The column is washed, first with 25 ml toluene-acetic acid (9:1) and then with 50 ml ether-hexane (3:1); the washes are discarded. Aflatoxins are eluted with 60 ml methylene chloride-acetone (9:1), and the eluate is evaporated to near-dryness on steam bath. The residue is dissolved in 1 ml methylene chloride and if necessary centrifuged to remove any particulate matter. The solution is transferred to a foil-lined screw cap and evaporated to dryness under a stream of nitrogen.

HPLC

The dry sample is dissolved in a suitable volume of elution solvent. HPLC standard (10 µl) is injected into the column and the chromatogram is recorded, followed by 10 µl of the sample extract. For confirmation 5 µl of standard and 5 µl of the sample extract are co-injected. The shape, symmetry, and retention times of the standards and sample mixture are indicative of the presence of aflatoxins.

TLC

Aliquots of sample extract and standards are spotted on an Adsorbosil-1 plate and developed for 40 minutes with anhydrous ethyl ether-methanol-water (96 + 3 + 1). Aflatoxins are quantitated either by visual comparison or by a fluorodensitometric scan. They can be recovered by extraction with methanol-10% NaCl solution (4:1).

REFERENCES

1. Dunmire, D. L. and Otto, S. E., "High Pressure Liquid Chromatographic Determination of Sugars in Various Products," *J. Assoc. Offic. Anal. Chem.* **62**:176 (1979).
2. Korbelak, T. and Bartlett, J. N., "The Separation and Identification of Four Synthetic Sweeteners by Thin-Layer Chromatography," *J. Chromatogr.* **41**:124 (1969).
3. Doner, L. W., White, J. W. and Phillips, J. G., "Gas-Liquid Chromatographic Test for Honey Adulteration by High-Fructose Corn Syrup," *J. Assoc. Offic. Anal. Chem.* **62**:186 (1979).
4. Kreiser, W. R. and Martin, R. A., "High Pressure Liquid Chromatographic Determination of Theobromine and Caffeine in Cocoa and Chocolate Products," *J. Assoc. Offic. Anal. Chem.* **61**:1424 (1978).
5. Verzele, M., Mussche, P. and Qureshi, S. A., "High-Performance Liquid Chromatographic Analysis of the Pungent Principles of Pepper and Pepper Extracts," *J. Chromatogr.* **172**:493 (1979).
6. Thompson, J. N. and Hatina, G., "Determination of Tocopherols and Tocotrienols in Foods and Tissues by High-Performance Liquid Chromatography," *J. Liquid Chromatogr.* **2**:327 (1979).
7. Leuenberger, U., Gauch, R. and Baumgartner, E., "Determination of Food Preservatives and Saccharin by High Performance Liquid Chromatography," *J. Chromatogr.* **173**:343 (1979).
8. Gosselé, J. A. W., "Gas Chromatographic Determination of Preservatives in Food," *J. Chromatogr.* **63**:429 (1971).
9. Gosselé, J. A. W., "Modified Thin-Layer Chromatographic Separation of Preservatives," *J. Chromatogr.* **63**:433 (1971).

10. Pelrović, S. E., Belin, B. E. and Vukajlovic, D. B., "Water-Soluble Vitamins," *Anal. Chem.* **40**:1007 (1968).

11. "Separation of Antioxidants by High-Pressure Liquid Chromatography," in: *Liquid Chromatography Report*, Du-Pont, Analytical Instruments Division, 1981.

12. Zakaria, N., Simpson, K., Brown, P. R. and Krstulovic, A., "Use of Reversed-Phase High-Performance Liquid Chromatographic Analysis for the Determination of Pro-vitamin A Carotenes in Tomatoes," *J. Chromatogr.* **176**:109 (1979).

13. Gross, J., "A Rapid Separation of Citrus Carotenoids by Thin-Layer Chromatography," *Chromatographie* **13**:572 (1980).

14. Martin, G. E., Tenenbaum, M., Alfonso, F. and Dyer, R. H., "High Pressure Liquid and Thin-Layer Chromatography of Synthetic Acid-Fast Dyes in Alcoholic Products," *J. Assoc. Offic. Anal. Chem.* **61**:908 (1978).

15. Sheppard, A. J., Newkirk, D. R., Hubburd, W. D. and Osgood, T., "Gas-Liquid Chro-matographic Determination of Cholesterol and Other Sterols in Foods," *J. Assoc. Offic. Anal. Chem.* **60**:1302 (1977).

16. Pons, W. A., "High-Pressure Liquid Chromatographic Determination of Aflatoxins in Corn," *J. Assoc. Offic. Anal. Chem.* **62**:586 (1979).

RECOMMENDED READING

Charlambous, G. (Ed.), *Liquid Chromatographic Analyses of Food and Beverages*, Vol. 2, Academic Press, 1979.

Clifford, A. J., "Nutrition: An Inviting Field to HPLC," in: *Advances in Chromatography* (J. G. Giddings, E. Grushka, J. Cazes and P. R. Brown, Eds.), Vol. 14, p. 1, Marcel Dekker, Inc., 1976.

Conacher, H. B. S. and Page, B. D., "Derivative Formation in the Chromatographic Analysis of Food Additives," *J. Chromatogr. Sci.* **17**:188 (1979).

Engelhardt, H., *High Performance Liquid Chromatography: Chemical Laboratory Practice*, Springer-Verlag, 1979.

Horváth, C. (Ed.), *Liquid Chromatography*, Academic Press, 1980.

Issaq, I. J. and Cutchin, W., "A Guide to Thin-Layer Chromatographic Systems for the Separation of Aflatoxins B_1, B_2, G_1 and G_2," *J. Liquid Chromatography* **4**:1087 (1981).

Lawrence, J. F. and Frei, R. W., *Chemical Derivatization in Liquid Chromatography*, Elsevier Publ. Co., 1976.

Official Methods of Analysis, 13th edition (Chap. 26: "Mycotoxins Methodology"), Associa-tion of Official Analytical Chemists, 1980.

Pryde, A. and Gilbert, M. T., *Applications of High Performance Liquid Chromatography*, Chapman and Hall, 1979.

Toxicants Occurring Naturally in Foods, National Academy of Sciences, Washington, D.C., 1966.

5
Forensic Analysis

INTRODUCTION

Forensic science is a broad field which encompasses all aspects of the application of scientific principles to the establishment of criminal guilt or innocence, including such specialities as pathology, psychiatry and jurisprudence. Criminalistics, a subdivision of forensic science, involves the collection and laboratory examination of physical evidence from the scene of a crime or a suspicious occurrence and court testimony on its significance in a particular case. Items submitted to a criminalistics laboratory might include a blood sample from a suspected drunken driver, a weapon or explosives obtained from the scene of a crime or a suspected forgery.

The increasing recognition of the investigative value of such evidence and its wide acceptance by the courts have created the need for further improvement of micro-chromatographic techniques for the examination of various organic compounds such as drugs, explosives and inks.

The following procedures describe the detection of drugs (alkaloids and cannabinoids), post-explosion residues and ball-point pen ink constituents using TLC, GLC and HPLC.

EXAMINATION OF BALL-POINT PEN INK CONSTITUENTS BY HPLC [1]

The examination of questioned documents by analysis of ink components is of major importance, especially in determining whether the said documents have been back-dated.

Over the past decade, the use of ball-point pens has greatly increased due to the improvements in ink formulations that have led to a performance similar to that of fountain-pen inks. Because of their increased use, ball-point inks are frequently involved when questioned documents are examined.

Ball-point pen inks are composed of two broad groups of compounds: the vehicles, which have an oil base, and the dyes.

Modern formulations use, almost exclusively, a glycol base with various resins. In addition, lipids are added as viscosity adjusters and dispersants, as well as small quantities of formaldehyde, phenol and thymol as preservatives. The dyes include triphenylamines, indolines and phthalocyanins. For such examinations HPLC is often quantitatively superior to TLC. Moreover, a UV detector can be used with HPLC to examine components of the vehicle that are not visible to the eye. Such compounds are not currently analyzed by TLC because they do not fluoresce or do not form visible spots with the usual TLC spray reagents.

Equipment and Materials

Column: HPLC tubing (25 cm long, 3 mm i.d.) packed with 10-μm silica-gel particles.
Mobile phase: a solution of 2% formamide in methanol; flow rate—0.5 ml/minute.
Detection: visible (580 nm) and UV (254 nm).

Procedure

Ten plugs from a written line are punched with a syringe needle and extracted with a few drops of 2% formamide in methanol. The ink extract (10 μl) is injected into the HPLC system. For direct analysis of vehicle components, punched-out dots from a written line are extracted with 2% isopropanol in heptane. Aliquots (10 μl) of the extract are injected into the HPLC system and eluted with 2% isopropanol in heptane at a flow rate of 0.5 ml/minute. Since the vehicle components are not colored, only UV detection (254 nm) is employed.

For blue inks an absorption wavelength of 580 nm in the visible region is used. Other wavelengths can be used for inks of other colors.

The assignment of particular peaks to the given dyes is based on matching the retention times with those of the standards.

The solvent system for extracting the ink from *felt-tipped* pens consists of dichloromethane-ethanol-formamide (89 + 10 + 1).

HPLC DETERMINATIONS OF THE MAJOR ALKALOIDS IN *PAPAVER SOMNIFERUM* [2]

Morphine was first isolated from opium in 1805. Since that time there has been no general agreement on a uniform method for quantitating

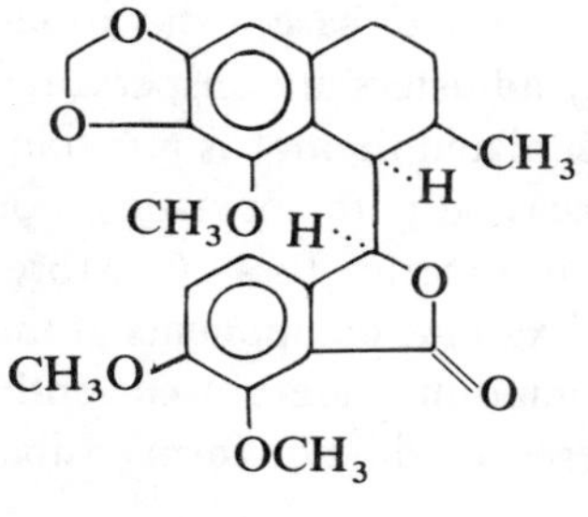

Narcotine

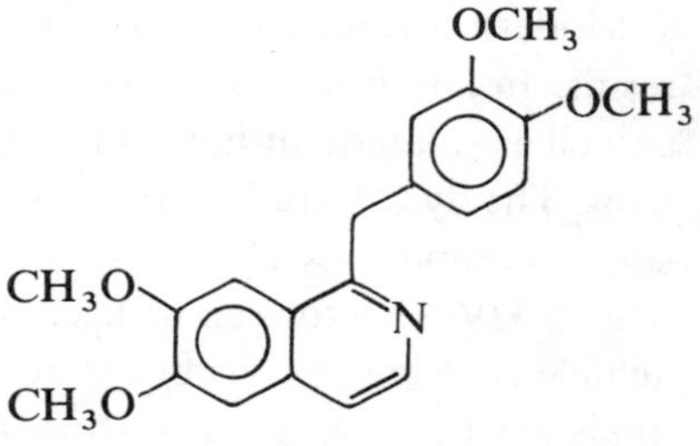

Papaverine

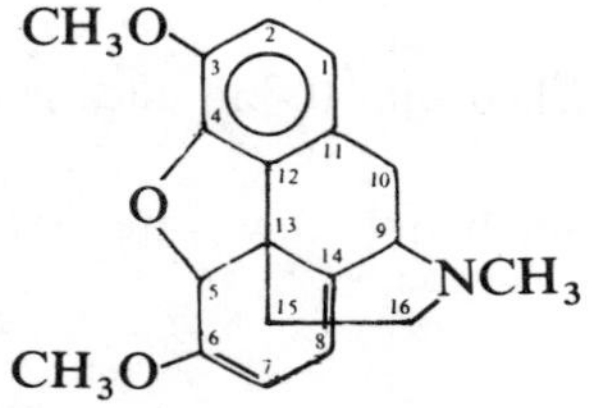

Thebaine

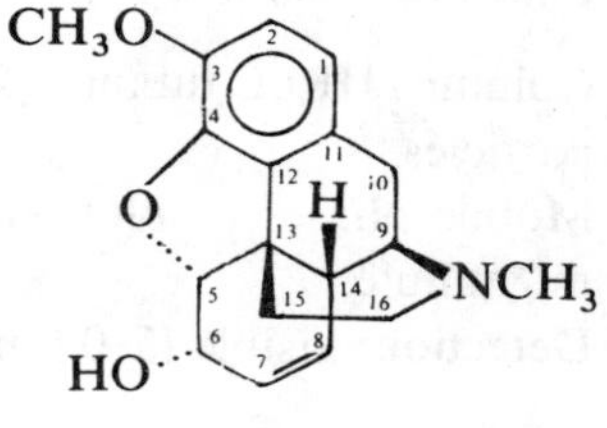

Codeine

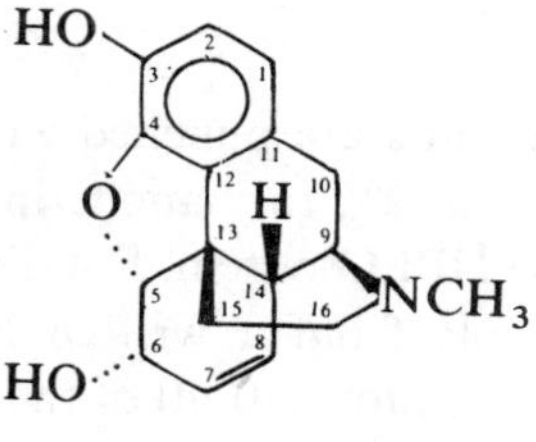

Morphine

morphine and different kinds of related alkaloids present in opium. Certain inherent problems are encountered, such as the limited solubility of morphine in organic solvents, the incomplete extraction of the alkaloids from the opium gum, and the adsorption of the alkaloids on various column packings during chromatographic separations. The following method is rapid and quantitative and is carried out on 10-μm porous silica-gel columns without pre-column purification.

Equipment and Materials

Column: stainless steel (30 cm long, 4 mm i.d.) packed with 10-μm silica gel. Mobile phase: *n*-hexane-methylene chloride-ethanol-diethylamine (300 + 30 + 40 + 0.5); flow rate—2.4 ml/minute at 1850 psi.

Detection: UV (285 nm).

Alkaloid standard solutions: prepared by dissolving 0.05 g narcotine, 0.01 g papaverine, 0.1 g thebaine, 0.2 g codeine and 0.4 g morphine in 100 ml ethanol.

Procedure

Capsular tissues of *Papaver somniferum* or *P. bracteatum* are pulverized with a ball mill. Fifteen milliliters of aqueous acetic acid (5 %) are added to 1 g of plant material in a stoppered 50-ml glass centrifuge tube, the contents are agitated thoroughly, and the tube is then placed in sonic bath for 30 minutes. The solution is then adjusted to pH 8.5 with a concentrated solution of ammonium hydroxide, followed by 20 ml of chloroform-isopropanol (3:1). After thorough agitation it is sonicated for 10 minutes. The emulsion is broken by centrifugation for 10 minutes at 2000 rpm, and the lower layer of chloroform-isopropanol is removed with a Pasteur pipette. The extraction is repeated three times. The extracts are combined and evaporated to dryness under vacuum. Absolute ethanol (1 ml) is added to the residue and the solution is kept in a sealed vial at $-10°C$.

The capsular extract is analyzed directly by HPLC; an aliquot of the extract mixed with the alkaloid standards is also analyzed for purposes of comparison. The order of elution of the alkaloids from the column is as follows: narcotine, papaverine, thebaine, codeine, morphine.

Since HPLC does not destroy the alkaloids, they can be recovered and stored for further studies.

SEPARATION AND IDENTIFICATION OF CANNABIS CONSTITUENTS BY TLC [3]

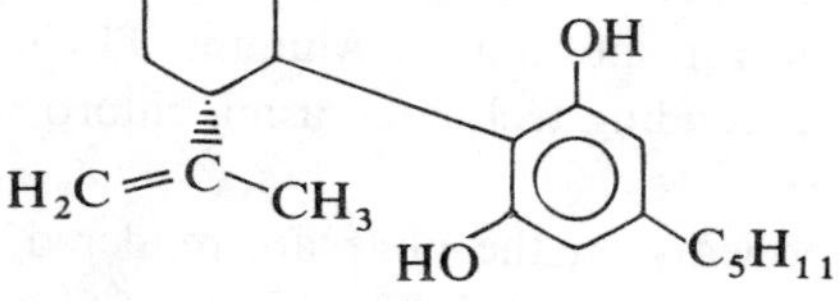

Cannabidiol 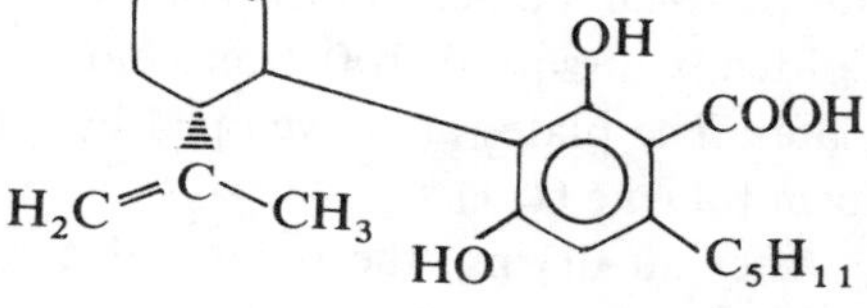   Cannabidiolic acid

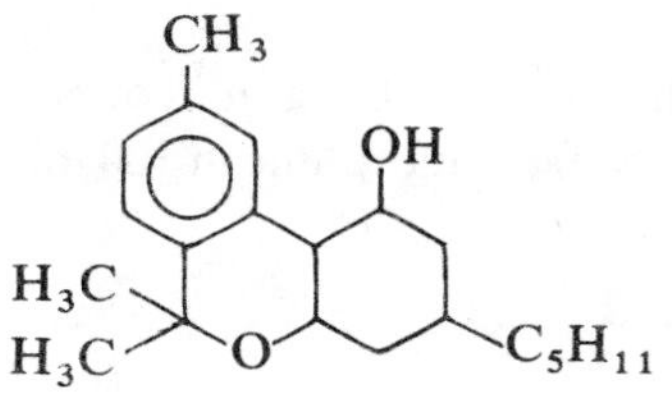

CH$_3$

OH

H$_3$C

H$_3$C   O   C$_5$H$_{11}$

Cannabinol

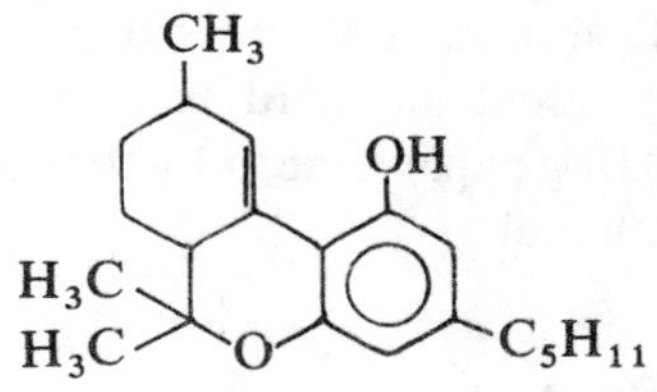

CH$_3$

OH

H$_3$C

H$_3$C   O   C$_5$H$_{11}$

Tetrahydrocannabinol I

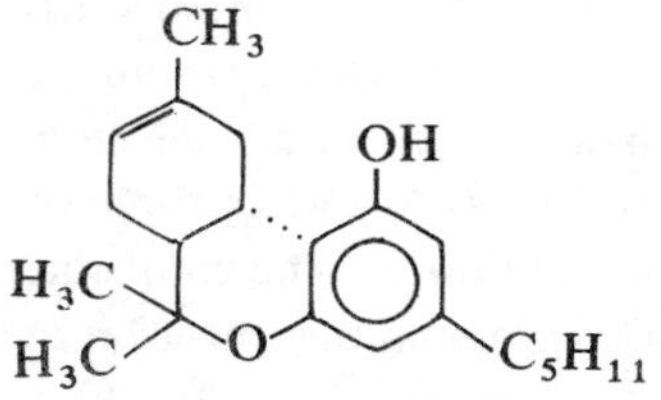

CH$_3$

OH

H$_3$C

H$_3$C   O   C$_5$H$_{11}$

Tetrahydrocannabinol II

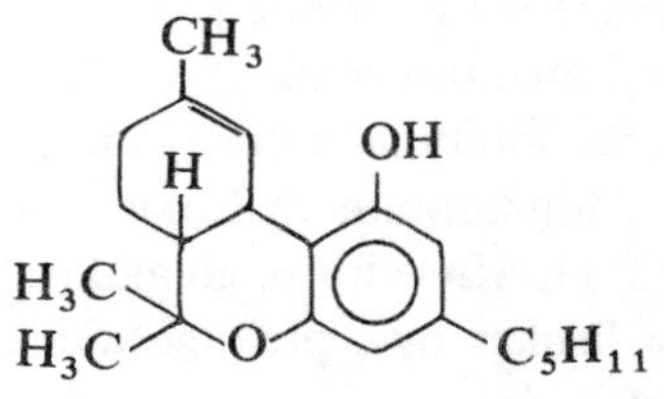

CH$_3$

H   OH

H$_3$C

H$_3$C   O   C$_5$H$_{11}$

Tetrahydrocannabinol III

The constant increase in the illicit trading of marijuana (hashish) in various forms has prompted the search for new microtechniques for its identification. Cannabis identification is usually based on macroscopic and microscopic examination of the destructive cannabis morphology and on color reactions.

In cases of samples where the morphology is absent or damaged, such as residues obtained from smoking pipes, other supporting techniques must be employed for positive identification. The following method for the separation of marijuana compounds is both rapid and sensitive.

Procedure

Dry powdered or sieved plant material or cannabis (0.1 g) is extracted with hot petroleum ether (5 ml). The extract is filtered and evaporated, and the residue is dissolved in 0.2 ml chloroform for spotting on Alumina TLC-plates. The plates are developed by the ascending technique using chloroform-toluene (1:1).

After air-drying, the separated components on the plate are rendered visible by UV light (254 nm) and by spraying with a 0.15% solution of fast Blue-BB salt, followed by 5% potassium hydroxide solution in methanol (see Table 8).

Table 8

Cannabis Constituent	R_f	Color of Spots
Cannabidiolic acid (CBDA)	0.8	Orange-pink
Tetrahydrocannabinol III	0.73	Grayish-violet
Tetrahydrocannabinol II	0.64	Blue-red
Tetrahydrocannabinol I	0.54	Scarlet
Cannabidiol (CBD)	0.44	Dark violet
Cannabinol (CBN)	0.27	Pink

If the sample consists of the residue from a smoking pipe, only the spots for cannabinol, cannabidiol, tetrahydrocannabinol II and two unidentified constituents (R_f 0.00 and 0.06) are obtained. The spots of CBDA and the two isomers tetrahydrocannabinol I and III are not obtained on the TLC-plate, probably due to the destruction of these constituents by the heat of the smoking pipe.

NINHYDRIN AS A SPRAY REAGENT FOR THE DETECTION OF SOME BASIC DRUGS ON THIN-LAYER CHROMATOGRAMS [4]

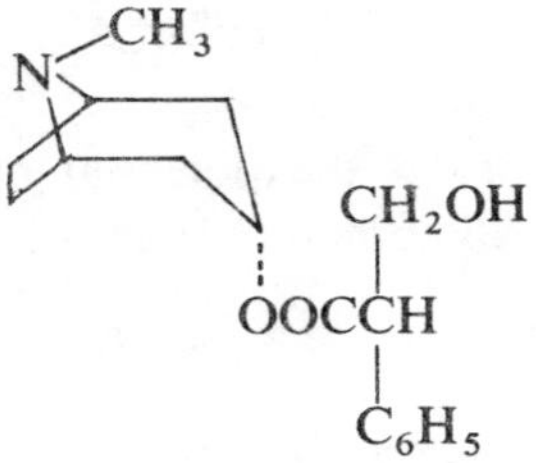

Atropine

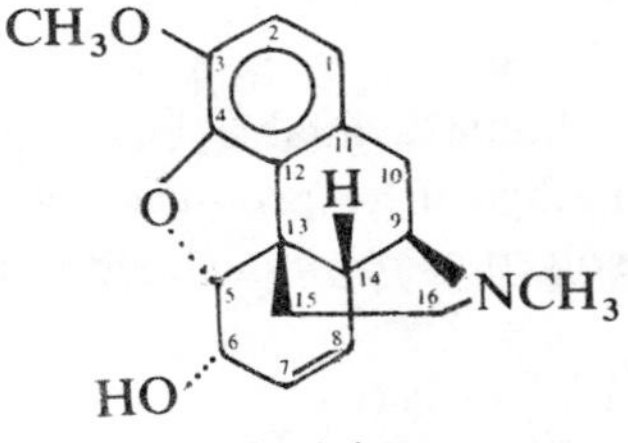

Codeine

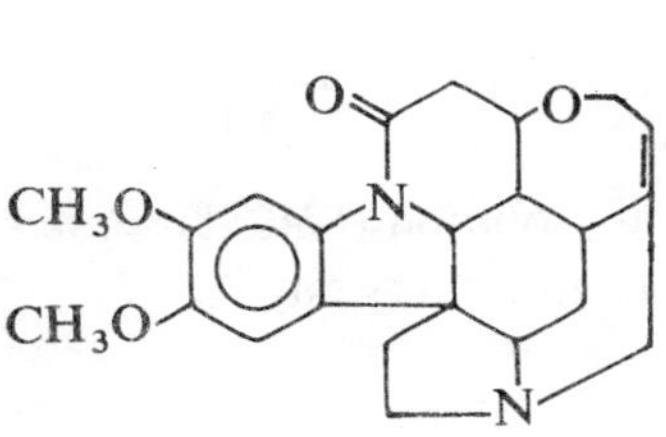

Brucine

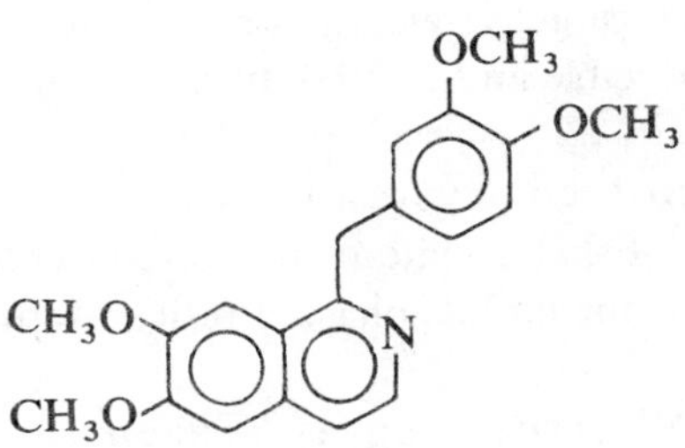

Papaverine

Strychnine   Procaine

TLC has become a regularly employed analytical tool for the separation and identification of drugs and other compounds. Among the reasons for its success are its sensitivity, economy, simplicity, high resolution and ability to test several unknowns and standards simultaneously.

The many reagents commonly used as a spray reagents for revealing basic drugs on TLC-plates include iodoplatinate, Dragendorff reagent, Marquis reagent, p-dimethylaminobenzaldehyde and bromocresol green. Usually these reagents produce one color or shades of the same color with basic drugs.

Ninhydrin, which is widely utilized for the detection of amino acids, can also be used to detect pharmaceutical compounds. Various colors are produced by this reagent on TLC-plates.

Equipment and Materials

Drugs
   Some available basic drugs are chosen to represent the potential of ninhydrin to produce color spots on TLC-plates. All compounds are dissolved as recommended in the Merck Index.

TLC-plates
   Merck TLC Plastic Rolls pre-coated with silica gel or silica-gel layers on alumina backing $60F_{254}$ (layer thickness 0.2 mm) are used. Self-coated silica gel G glass plates were found to be unsuitable as they give unreproducible and often diffuse patterns.

Solvent system
   Ethyl acetate-methanol-concentrated ammonia solution (17:2:1) is used as the mobile phase. Fresh solvent has to be prepared every day.

Ninhydrin spray reagent
   A 10% solution of ninhydrin in 95% ethanol is used.

Procedure

After development the plates are dried in an air oven at 100°C for 3–4 minutes to remove ammonia. Incomplete removal of ammonia from the plates prior to spraying with the ninhydrin reagent causes a brownish background which tends to mask the color of the spot. The plates are cooled to room temperature and sprayed with the ninhydrin reagent. Gloves must be worn when this reagent is sprayed.

The plates are heated at 120°C for 30 minutes and the colors of the spots and their R_f values are recorded (see Table 9).

Table 9

Drug	R_f	Color of Spots
Atropine sulfate	0.17	Blue
Brucine	0.23	Pink
Strychnine hydrochloride	0.35	Pink
Codeine phosphate	0.41	Blue
Papaverine hydrochloride	0.64	Orange
Procaine hydrochloride	0.65	Pink-violet

IDENTIFICATION OF POST-EXPLOSION RESIDUES BY HPLC AND TLC [5]

Chromatographic (TLC, GLC, HPLC), spectroscopic (IR, UV) and chemical tests are employed for identifying explosives in post-explosion cases. Most explosives are classified into the following groups:

(a) standard military high explosives, such as 2,4,6-trinitrotoluene (TNT), pentaerythrytol tetranitrate (PETN) and 1,3,5-trinitro-1,3,5-triazacyclohexane (RDX, Cyclonite, Hexogen);

(b) standard military low explosives (propellants), such as smokeless powder [e.g., a mixture of nitrocellulose and nitroglycerine (NG)] collected from ammunition;

(c) dynamites—mixture of NG and ethyleneglycoldinitrate (EGDN) or ammonium nitrate (NH_4NO_3);

(d) home-made mixtures containing an oxidizing agent (nitrates, chlorates, permanganates) and a reducing agent (aluminum, charcoal, sulfur, sugar).

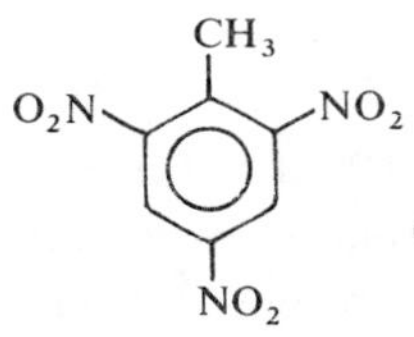

TNT 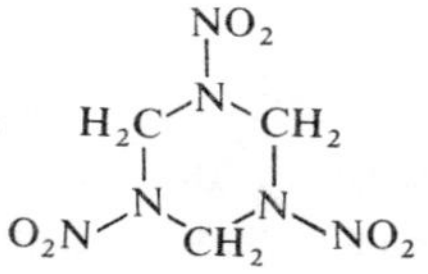   RDX 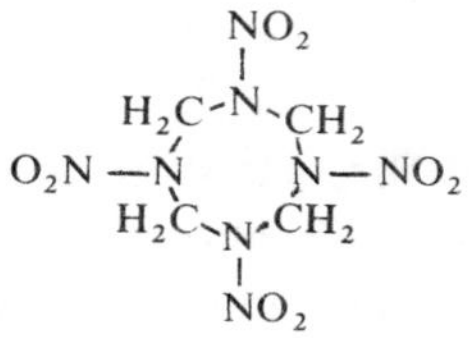   HMX 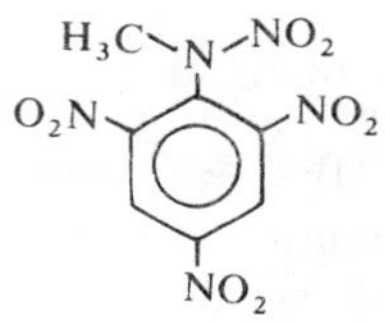   Tetryl

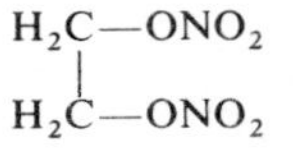

EGDN 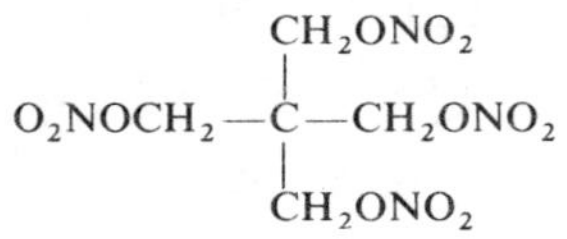   PETN 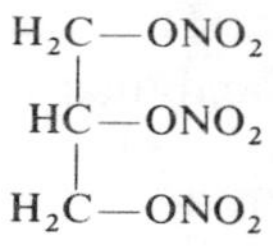   NG

TLC and HPLC techniques are used for the identification of explosives of groups (a), (b) and (c).

Equipment and Materials

HPLC

Column: stainless steel (25 cm long, 4.6 mm i.d.) packed with 10μ bonded polyglosil $\cdot$ $N(CH_3)_2$.

Mobile phase: solution of 20% isopropanol in hexane; flow rate—60 ml/hour.

Detection: UV (220 nm).

Explosives standards: prepared in acetone solutions.

TLC

Potassium hydroxide solution: prepared by dissolving 2 g KOH in 100 ml of 95% ethanol.

Modified Griess reagent: prepared by mixing equal volumes of the following freshly prepared solutions: solution 1—sulphanylamide (8 g) dissolved in 85% phosphoric acid (A.R. NO_2^--free) (10 ml) and made up to 100 ml with distilled water; solution 2—N-(1-naphthyl)ethylenediamine $\cdot$ 2HCl (0.55 g) dissolved in 85% phosphoric acid (A.R. NO_2^--free) (10 ml) and made up to 100 ml with distilled water.

Ethylenediamine-dimethylsulfoxide (EDA-DMSO) is prepared by mixing equal volumes of EDA and DMSO.

Treatment of Collected Debris

The debris are extracted with hot pure acetone. The solvent is removed by evaporation. If quantities are large and the analysis must be performed quickly, a rotating evaporator may be used.

Analytical Procedure for Organic Extract

TLC

A few milligrams (2 to 6) of the evaporation residue dissolved in 1 ml acetone are spotted on silica-gel plate, as well as 1–2 µl of standard solutions (1–5 µg/µl acetone) of common explosives. After development with trichloroethylene-acetone mixture (4:1), the plate is sprayed with KOH solution followed by heating the plate at 100–110°C for 10 minutes. The modified Griess reagent is then applied. The alkaline solution not only detects the nitroaromatic compounds, but also hydrolyzes the nitrate esters (PETN, NG, EGDN) and nitramines (RDX, HMX, Tetryl) to form (especially upon heating) the NO_2^- ions needed for the Griess reaction. The R_f values in decreasing order are: TNT (0.59); Tetryl (0.40); EGDN (0.47); NG (0.47); PETN (0.55); RDX (0.15); HMX (0.05). The first three compounds are colored brown-red or orange-brown when sprayed with KOH. These compounds can also be detected by EDA-DMSO, a color reagent for polynitroaromatic compounds.

HPLC

The following are the retention-time values: TNT—0.6 minutes; NG — 6.7 minutes; PETN—9.4 minutes; RDX—11.8 minutes.

REFERENCES

1. Colwell, L. F. and Karger, B. L., "Ball-Point Pen Ink Examination by High Performance Liquid Chromatography," *J. Assoc. Offic. Anal. Chem.* **60**:613 (1977).
2. Vincent, P. G. and Engelke, B. F., "High Pressure Liquid Chromatographic Determination of the Five Major Alkaloids in *Papaver somniferum* L. and Thebaine in *Papaver bracteatum* Lindl. Capsular Tissue," *J. Assoc. Offic. Anal. Chem.* **62**:310 (1979).
3. Tewari, S. N., Harpalani, S. P. and Sharma, S. C., "Separation and Identification of the Constituents of Hashish (*Cannabis indica* Linn.) by Thin-Layer Chromatography and Its Application in Forensic Analysis," *Chromatographia* **7**:205 (1974).

4. Dutt, M. C. and Poh, T. T., "Use of Ninhydrin as a Spray Reagent for the Detection of Some Basic Drugs in Thin-Layer Chromatograms," *J. Chromatogr.* **195**:133 (1980).
5. Glattstein, B. and Kraus, S., "Identification of Post-Explosion Residues by HPLC and TLC," Police Anal. Lab. Jerusalem, private communication.

RECOMMENDED READING

Davis, G. (Ed.), *Forensic Science*, Amer. Chem. Soc. Series 13, 1975.
Macek, K. and Haller, A., "Chromatography of Drugs," in: *Chromatography* (E. Hoffmann, Ed.), pp. 675–713, Van Nostrand Co., 1975.
Mechoulam, R. (Ed.), *Marijuana; Chemistry, Pharmacology, Metabolism and Clinical Effects*, Academic Press, 1973.
Oliver, J. S. (Ed.), *Forensic Toxicology*, Croom-Helm, London, 1980.
Smith, R. N., "Chromatography in Forensic Science," in: *Developments in Chromatography*, Vol. 1 (C. E. H. Knapman, Ed.), pp. 201–239, Applied Science Publishers, London, 1978.
Stahl, E., *Drug Analysis by Chromatography and Microscopy*, Ann Arbor Science Publishers, 1973.
Vanden Heuvel, W. J. and Zacchei, A. G., "Gas-Liquid Chromatography in Drug Analysis," in: *Advances in Chromatography* (J. G. Giddings, E. Grushka, J. Cazes and P. R. Brown, Eds.), Vol. 15, p. 199, Marcel Dekker, Inc., 1977.

Subject Index

Notes

Notes

Notes

Notes